智能制造英语

主　编：徐利君　黎　霞　陈利红

副主编：黄　虹　朱永波　申　俊　陈开银

北京理工大学出版社
BEIJING INSTITUTE OF TECHNOLOGY PRESS

图书在版编目（CIP）数据

智能制造英语 / 徐利君，黎霞，陈利红主编．
北京：北京理工大学出版社，2024. 12（2025. 1 重印）.
ISBN 978 - 7 - 5763 - 4008 - 2

Ⅰ. TH166

中国国家版本馆 CIP 数据核字第 2024NZ4836 号

责任编辑：芈　岚　　　文案编辑：芈　岚
责任校对：刘亚男　　　责任印制：施胜娟

出版发行 / 北京理工大学出版社有限责任公司
社　　址 / 北京市丰台区四合庄路 6 号
邮　　编 / 100070
电　　话 /（010）68914026（教材售后服务热线）
　　　　　（010）63726648（课件资源服务热线）
网　　址 / http://www. bitpress. com. cn

版 印 次 / 2025 年 1 月第 1 版第 2 次印刷
印　　刷 / 涿州市京南印刷厂
开　　本 / 787 mm × 1092 mm　1/16
印　　张 / 11. 25
字　　数 / 200 千字
定　　价 / 66. 00 元

前 言

随着全球工业的迅猛发展和技术的持续创新，智能制造已成为我国推动产业结构升级、应对市场需求变化、提升国际竞争力的核心战略之一。智能制造不仅展示了我国在科技领域的卓越成就，更是新时代中国工匠精神的传承与发扬。党的二十大报告提出“推进新型工业化，加快建设制造强国”，并明确“推动制造业高端化、智能化、绿色化发展”。在此战略背景下，高等职业院校肩负着培养高素质技能型人才的重要使命，助力中国制造向中国“智造”的全面转型。

本教材依据《高等职业教育专科英语课程标准（2021 年版）》编写，属于课标中的拓展模块——职业提升英语，旨在为装备制造大类专业的学生提供一个专业学习英语的平台。为实现基础英语学习和专业英语学习的有机衔接，“智能制造英语”课程可设置在基础英语学习之后或专业英语学习之前，形成基础英语、智能制造英语和专业英语灵活搭配的课程体系。

本教材在编写过程中，注重核心素养的培养，力求将学生培养成为具备专业技能、国际视野和文化自信的新时代职业人才。

一、教材特色

1. 传递职业责任，培养文化自信。本教材融合了中国制造领域的最新成果与技术，提供了丰富的代表性案例，不仅引导学生深刻理解并认同国家在科技领域的进步，而且在他们心中植入职业责任感的种子。本教材通过与教学实践相结合，唤起学生参与全球制造业变革的使命感，并以此为契机，拓宽学生的国际视野，厚植文化自信，确保他们在未来能够以坚定的文化自信和强烈的责任感，积极推动制造业的创新与发展。

2. 塑造职业品格，弘扬工匠精神。本教材为项目式教材，秉承“知行合一”的教育理念，以学生为中心，推崇探究式学习方式，致力于在实践中培养学生的职业品格。

本教材围绕智能制造的关键技术领域，巧妙设计了一系列单元项目，将工业文化的精髓与英语教学有机结合。通过这些项目，学生不仅能够锻炼自己的专业技能，还能够在团队合作中培养协作精神；同时，能在解决实际问题的过程中，深刻领悟和践行工匠精神，从而塑造出精益求精、追求卓越的职业品格。

3. 贴近职业需求，强化英语应用。本教材的每个单元均设有项目拓展环节，以进一步强化学生的英语语言技能与实际职业需求之间的联系。该环节设计了多个智能制造行业的实际任务，涵盖操作工业机器人、操作数控机床、监控大数据平台、灯塔工厂调研、操作虚拟生产线、分析制造业出海等主题，这些任务有助于学生在真实职业环境中更有效地应用英语，以适应不断变化的行业需求。

二、项目说明

本教材作为项目式教材，围绕智能制造中的生产、连接、数据、集成、创新与转型六大核心领域顺序展开，每个单元均采用“项目线”和“语言线”并行的结构。项目成果分成两种形式：制作表现类成果和解释说明类成果，具体包括海报设计、视频制作、宣传册编写、角色扮演、PPT 演示、演讲等，符合英语学科的特性。

在项目实施过程中，学生将在真实的职业情境中综合运用听、说、读、看、写、译等语言技能，并在实践环节锻炼自己的操作能力。整个项目的实施遵循“由大到小、逐步聚焦”的原则，帮助学生从宽泛的知识理解逐步深入到具体的技能应用，通过多个环节逐步提升综合能力，并完成项目成果的展示。

项目完成后，学生将依据单元项目评价量规进行反思、评价、修正，以进一步深化对学习内容的理解，确保核心素养的全面提升。

三、教材结构

部分	说明
Unit Introduction	帮助学生明确项目的总体目标和方向
Getting Ready	激活学生已有的相关知识或经历，提升学生对项目主题的兴趣，并为其提供更多完成项目所需的语言和背景知识
Starting Up	引导学生了解项目实施的背景和意义，剖析项目中需要解决的核心问题，帮助学生为后续步骤做好准备
Investigating	引入行业相关的先进理念与技术，帮助学生将理论与实际相结合，理解项目的背景与挑战
Researching	通过研究真实案例，提升学生的行业理解力和研究能力，为项目实施提供数据支持和理论依据

续表

部分	说明
Building and Presenting	这是项目的实施与成果展示阶段，学生按照规定的步骤执行项目，并通过成果展示提升表达和沟通能力
Assessing and Reflecting	为学生提供自评、互评和反思的机会，帮助他们总结学习成果，分析项目中的挑战，并通过公开展示项目成果，促进学生的持续改进和深入学习
Extending	这是项目拓展环节，以项目的延伸内容为载体，帮助学生在实际工作场景中应用所学技能，进一步强化实践能力

本教材由湖南机电职业技术学院的英语教师以及信息工程学院、电气工程学院、机械工程学院的专业教师，还有湖南宇环智能装备有限公司技术部高级工程师陈开银先生共同编写完成。具体分工如下：徐利君负责设计编写理念和各单元内容，并进行修改统稿；黎霞负责设计搭建项目并校稿；陈利红负责设计制作习题；黄虹负责设计制作图表；朱永波、申俊负责提供专业素材；陈开银负责提供企业案例。本教材编写过程中，得到了湖南机电职业技术学院王建红、郭蓉菊、李明霞、易艺、彭俊、王茜、史炜炜等老师的热情帮助，在此一并表示衷心感谢。由于编者水平有限，书中难免存在不足之处，敬请同行和读者批评指正，以便再版时加以改进。

《智能制造英语》编写组

Contents

Contents

Researching	Building and Presenting	Assessing and Reflecting	Extending
Cases: The main application of industrial robots in manufacturing in China	◆ **Building**: Create an informative poster for your industrial robot. ◆ **Presenting**: Present the project in class as a group.	◆ **Assessing**: Evaluate the project jointly by teachers and students. ◆ **Reflecting**: Reflect by answering questions.	**Practical Operation**: Industrial robotic arms
Cases: The application of connected high-end CNC machines in manufacturing in China	◆ **Building**: Create a promotional video for your CNC machine. ◆ **Presenting**: Present the project in class as a group.	◆ **Assessing**: Evaluate the project jointly by teachers and students. ◆ **Reflecting**: Reflect by answering questions.	**Practical Operation**: High – end CNC machines
Cases: The main application of big data platforms in manufacturing in China	◆ **Building**: Create a brochure for your big data platform. ◆ **Presenting**: Present the project in class as a group.	◆ **Assessing**: Evaluate the project jointly by teachers and students. ◆ **Reflecting**: Reflect by answering questions.	**Practical Operation**: Industrial cloud platforms
Cases: China's lighthouse factories	◆ **Building**: Create a role play about a lighthouse factory tour. ◆ **Presenting**: Present the project in class as a group.	◆ **Assessing**: Evaluate the project jointly by teachers and students. ◆ **Reflecting**: Reflect by answering questions.	**Practical Operation**: Lighthouse factory reports
Cases: AI innovations transforming manufacturing in China	◆ **Building and Presenting** Make a PPT presentation for your innovative applications.	◆ **Assessing**: Evaluate the project jointly by teachers and students. ◆ **Reflecting**: Reflect by answering questions.	**Practical Operation**: Virtual production lines
Cases: Digital transformation in manufacturing in China	◆ **Building and presenting** Give a public speech on your digital transformation.	◆ **Assessing**: Evaluate the project jointly by teachers and students. ◆ **Reflecting**: Reflect by answering questions.	**Practical Operation**: Chinese manufacturing going global

Unit 1
Production in Intelligent Manufacturing
智能制造生产

Unit Introduction

Driving Questions
驱动问题

With the rapid integration of digital technology and manufacturing, China's industry is advancing along the high-end value chain, transitioning from "Made in China" to "Intelligent Manufacturing(IM)". As we delve into the future of production, where humans and machines collaborate seamlessly, we are not only redefining possibilities but also pioneering a new era of industrial derelopment. What does "production in intelligent manufacturing" entail? As the core equipment on a production line, how are industrial robots revolutionizing the production process?

随着数字技术与制造业的快速融合,中国工业正向着高端价值链迈进,从"中国制造"转型为"智能制造"。未来的生产中,人机协作会变得更加和谐,我们不仅在不断刷新极限,也在开创一个工业发展的新纪元。那么,智能制造生产是什么?作为生产线上的核心设备,工业机器人是如何革新生产流程的?

Project Overview
项目概述

In this unit, you will begin by exploring the concept of intelligent manufacturing, understanding its principles and significance in the modern industrial landscape. After that, you will move on to the specific concept of intelligent manufacturing production, focusing on its integration with various technologies. Subsequently, you will learn about the concept of industrial robots and study practical application cases of their application in China's manufacturing industry. Next, working in groups, you will research and create an informative poster for ABC Company's industrial robots. Finally, you will present your poster to the class, discussing the practical significance and potential of industrial robots in advancing intelligent manufacturing production.

在本单元,首先要学习智能制造的概念,了解其在现代工业领域中的基本原理和重要性。然后是深入学习智能制造生产的具体概念,重点了解智能生产线如何集合各种技术;学习工业机器人的概念,研究中国制造业中工业机器人的实际应用案例。随后,小组合作为 ABC 公司的工业机器人制作一份宣传海报。最后,在班级展示海报并讨论工业机器人在推进智能制造生产方面的实际意义和潜力。

Project Extension

项目拓展

After completing the main project, you'll engage in a hands-on extension activity to deepen your understanding of industrial robots. This will involve learning how to operate a robotic arm to perform simple tasks.

完成主要项目后,将进行实操拓展,操作机械手臂执行简单的任务,以进一步巩固对工业机器人的理解。

Learning Objectives

学习目标

This unit is intended to help you:

1. have a general idea of intelligent manufacturing;
2. explore the structure and benefits of intelligent manufacturing production lines;
3. analyze the main applications of industrial robots in simple English;
4. create and present an English poster about industrial robots;
5. operate a robotic arm;
6. develop a sense of pride in "Made in China".

本单元旨在帮助你:

1. 了解智能制造;
2. 探讨智能制造生产线的结构和优势;
3. 用英语简单分析工业机器人的主要应用;
4. 制作并展示工业机器人的英文海报;
5. 操作机械手臂;
6. 培养对"中国制造"的自豪感。

Getting Ready

► What is a poster, and what are the linguistic features of its text? Please complete the tasks below to find the answer. Fig. 1. 1 is a sample poster.

什么是海报？海报文本有什么特点？请完成以下任务，找到答案。图 1. 1 是一张海报示例。

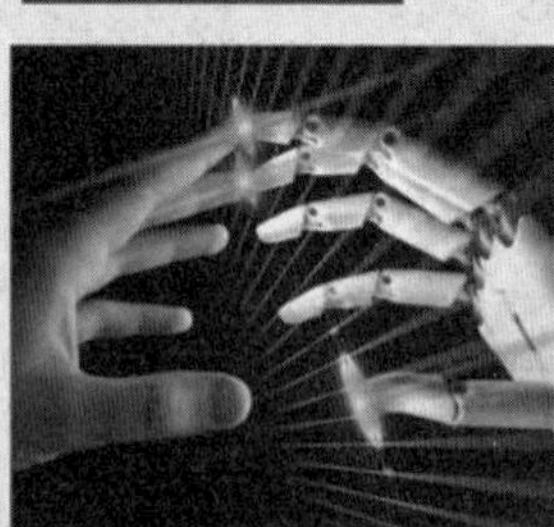

Fig. 1. 1　A poster created by STEM Innovator and Artist Shreya Pal

Task 1 *Listen to the passage about posters and complete the sentences with the words you hear.*

Posters are large sheets of 1. ________ that people put up in 2. ________ places to promote something or on walls as 3. ________. Most posters combine both text and 4. ________, though some might be purely or entirely written. The main goal of a poster is to grab 5. ________ and provide information. Posters serve many different purposes. They are commonly used by advertisers to promote 6. ________, musicians, films and more. They are also 7. ________ for propagandists and others who want to spread a message. Additionally, posters are often used to reproduce famous artworks, offering a cheaper alternative to the original pieces. In the past, many posters were created to 8. ________ products, and this continues today.

Task 2 *Learn some linguistic features of the poster text.*

2.1 简短句(Short sentences)

特点:海报通常使用简短的句子,以便迅速传递关键信息。

示例:

This new machine can double the rates of production.

↓

This machine doubles production rates.

2.2 省略句(Elliptical sentences)

特点:将句中一个或多个通常必需的语法成分省略,以便在海报有限的空间内提供必要信息。

示例:

The product is designed in China and manufactured with great precision.

↓

Designed in China, manufactured with precision.

2.3 缩写(Abbreviations)

特点:将一个较长的词或短语缩写成较短的形式,如简单的字母组合、首字母缩略词以及截短词,可用于需要节约空间的海报文本中。

示例：

information → info.
statistics → stats.
promotion → promo.
United States of America → USA
National Aeronautics and Space Administration → NASA

Task 3 *Rewrite the following sentences according to the requirements.*

1. Original: Our robots are designed to reduce human error.

   Short: ____.

2. Original: The safety of our employees is our top priority.

   Short: ____.

3. Original: Our designs are driven by the need for energy efficiency.

   Short: ____.

4. Original: Our company guarantees to deliver on time for every order.

   Short: ____.

5. Original: The battery of this phone makes it last longer.

   Short: ____.

6. Original: This equipment is built for both speed and precision.

   Elliptical: ____.

7. Original: Our products come with a guarantee of quality and assurance of customer satisfaction.

   Elliptical: ____.

8. Original: Our products are designed to last.

   Elliptical: ____.

9. Original: This offer is valid until December 31st.

   Elliptical: ____.

10. Original: The book was written by a Nobel Prize winner.

    Elliptical: ____.

11. Original: Our computer numerical control machines provide precise manufacturing solutions.

Abbreviated: ______________________________.

12. Original: Check out the specifications of our new model.

Abbreviated: ______________________________.

13. Original: Receive notifications for research and development.

Abbreviated: ______________________________.

14. Original: We use eco-friendly technology for a greener tomorrow.

Abbreviated: ______________________________.

15. Original: Demonstrations available upon request.

Abbreviated: ______________________________.

Starting up

▶ Intelligent manufacturing represents a new approach to production that is reshaping the industry. But what exactly is intelligent manufacturing? In this section, we'll learn about its core concepts, key components, driving forces and the value it adds to modern manufacturing.

智能制造是一种正在改变工业生产方式的新模式。那么,究竟什么是智能制造?在本部分中,我们将学习其核心概念、关键组成部分、驱动力以及它为现代制造业带来的价值。

Introduction to Intelligent Manufacturing

An intelligent manufacturing system is a human-machine integrated intelligent system, composed of intelligent machines and human experts. The development of intelligent manufacturing originates from research in artificial intelligence (AI), including intelligent manufacturing technologies and systems.

These systems are capable of carrying out intelligent activities in the manufacturing process, such as analysis, reasoning, judgment, conception, and decision-making. Additionally, they can continuously enhance their knowledge base through independent learning as well as collecting and interpreting both environmental and self-related information. They can also judge and plan their own actions, making them highly adaptive and autonomous. Through the cooperation of humans and intelligent machines, human expertise is extended, making manufacturing automation more flexible, intelligent, and highly integrated.

Key Components

Intelligent manufacturing is characterized by several key components.

- Connectivity: Ensures seamless communication between devices and systems.
- Automation: Increases efficiency by minimizing human intervention in repetitive tasks.
- Advanced data management and analytics: Enhance decision-making processes through data-driven insights.
- Connected ecosystems: Involve integrating different stakeholders, systems, and processes within the manufacturing environment.
- Innovative business strategies: Encourage adopting new business models and strategies to stay competitive.
- Agile operating models: Promote flexibility and adaptability in operations.
- Engaged, connected employees: Ensure that employees are well-integrated into the intelligent manufacturing framework and can interact effectively with automated systems.

These components collectively contribute to building a more intelligent and interconnected organization.

Driving Forces

Intelligent manufacturing is driven by various factors. Product cycles are becoming shorter, increasing the pressure to improve efficiency and sustainability. Additionally, the industry faces challenges related to workforce skills. These pressures and challenges are significant roadblocks to staying competitive, making intelligent manufacturing essential to thrive in the market.

The Value

The value of intelligent manufacturing is clear. Organizations that implement intelligent initiatives can achieve significant efficiency gains. This improvement underscores the power and necessity of intelligent manufacturing in modern industry.

In summary, intelligent manufacturing is a transformative approach that combines advanced technology, innovative strategies, and human expertise to enhance efficiency, sustainability, and competitiveness in the industry. This integrated approach is essential for modern industry to stay ahead in a rapidly evolving market.

New Words and Phrases

intelligent[ɪnˈtɛlɪdʒənt] *adj.* 智能的

manufacturing[ˌmænjʊˈfæktʃərɪŋ] *n.* 制造

judgment[ˈdʒʌdʒmənt] *n.* 判断

conception[kənˈsepʃ(ə)n] *n.* 构思

enhance[ɪnˈhæns] *v.* 增强

connectivity[ˌkənekˈtɪvəti] *n.* 连接(度),连接性

automation[ˌɔːtəˈmeɪʃ(ə)n] *n.* 自动化

ecosystem[ˈiːkəʊsɪstəm] *n.* 生态系统

innovative[ˈɪnəveɪtɪv] *adj.* 创新的,革新的

agile[ˈædʒaɪl] *adj.* 灵活的,敏捷的

adaptive[əˈdæptɪv] *adj.* 有适应能力的

autonomous[ɔːˈtɒnəməs] *adj.* 自主的

sustainability[səsˌteɪnəˈbɪlɪti] *n.* 可持续性

competitiveness[kəmˈpetətɪvnəs] *n.* 竞争力

initiative[ɪˈnɪʃətɪv] *n.* 倡议,新方案

efficiency[ɪˈfɪʃ(ə)nsi] *n.* 效率,功效

factor[ˈfæktə] *n.* 因素

thrive[θraɪv] *v.* 兴盛

underscore[ˌʌndərˈskɔːr] *v.* 强调

be composed of　由……组成

originate from　起源

be capable of　能够

carry out　执行

contribute to　有助于

be essential for　必不可少的

Technical Terms

human expertise　人类专业知识

artificial intelligence (AI)　人工智能

advanced data management and analytics　高级数据管理和分析

business strategies　商业战略

operating models　运营模式

product cycle　产品周期

workforce skills　劳动力技能

Task 4 *According to the passage, are the following statements true (T) or false (F)?*

(　　)1. Intelligent manufacturing systems consist of intelligent machines without human involvement.

(　　)2. Intelligent manufacturing systems are highly adaptive and autonomous.

(　　)3. Advanced data management and analytics play a crucial role in enhancing decision-making processes in intelligent manufacturing systems.

(　　)4. Engaged, connected employees are not the key component of intelligent manufacturing.

(　　)5. Agile operating models in intelligent manufacturing promote rigidity and inflexibility in operations.

(　　)6. One of the driving forces behind intelligent manufacturing is the pressure to improve efficiency and sustainability.

(　　)7. The value of intelligent manufacturing primarily lies in reducing costs, rather than improving efficiency and sustainability.

(　　)8. Intelligent manufacturing is a comprehensive approach that combines advanced technologies, innovative strategies, and human expertise.

Task 5 *Read the passage about the intelligent manufacturing, and answer the following questions.*

1. What is an intelligent manufacturing system?

____

2. List the key components of intelligent manufacturing.

____

3. What is the role of "connected ecosystems" in intelligent manufacturing?

____

4. What are the main drivers of intelligent manufacturing?

____

5. How does intelligent manufacturing help modern industries stay ahead in a rapidly evolving market?

____

Task 6 *Match the English expressions in Column A with their Chinese meanings in Column B.*

Column A	Column B
1. intelligent manufacturing system	A. 互联生态系统
2. artificial intelligence	B. 产品周期
3. data management and analytics	C. 智能制造系统
4. business strategies	D. 商业战略
5. operating models	E. 劳动力技能
6. product cycle	F. 运营模式
7. workforce skills	G. 人工智能
8. connected ecosystem	H. 数据管理和分析

Task 7 *Make six sentences with words from the following boxes.*

manufacturing **automation** **connectivity** **sustainability**

initiative **competitiveness** **efficiency** **product cycle**

Example: The **manufacturing** industry is one of the backbone industries in China.

1. ______________________________.
2. ______________________________.
3. ______________________________.
4. ______________________________.
5. ______________________________.
6. ______________________________.

Task 8 *Translate the following sentences into English using the words and expressions given.*

1. 工匠精神的精髓在于对细节的关注和对品质的追求。(be composed of)

____.

2. 灯笼起源于东汉,最初主要用于照明。(originate from)

____.

3. 庆祝中国节日有助于理解传统文化。(contribute to)

____.

4. 科技创新取得了一批重大成果。(innovative)

____.

5. 一旦作出决定,就要坚决执行。(carry out)

____.

Investigating

► The production process consists of a sequence of operations in a factory that work together to form a production line. In this section, we'll learn about the basics of an intelligent manufacturing production line.

生产过程由工厂中的一系列工序组成,这些工序共同形成了生产线。在本部分,我们将学习智能制造生产线的基本特点。

Intelligent Manufacturing Production Line

An intelligent manufacturing production line is a highly automated arrangement that utilizes intelligent manufacturing technology to optimize the production process.

It covers three levels of automated equipment, digital workshops, and intelligent factories and encompasses six key aspects of intelligent manufacturing: intelligent management, intelligent monitoring, intelligent processing, intelligent assembly, intelligent inspection, and intelligent logistics.

The production line integrates common technologies including digitalization, automation, informatization, and intelligentization.

This setup enables eight types of connections within the industrial internet, facilitating interactions among intelligent factories, factory control systems, work-in-progress (WIP) items, intelligent machines, industrial cloud platforms (including management software), and collaboration platforms, as well as smart products.

An intelligent manufacturingproduction line has the following characteristics.

➤ High integration

In the future, production lines will integrate hardware devices, software systems, and data platforms. Connected devices will form an automated network linked with manufacturing execution systems (MES) and enterprise resource planning (ERP) systems, achieving visualization, control, and optimization of the production process.

➤ Enhanced intelligence

With the help of AI and machine learning, intelligent manufacturing production lines possess self-learning and adaptive capabilities. They monitor equipment and product quality in real-time, adjusting production processes through data analysis to achieve intelligent management. Intelligent inspection and warehousing technologies further improve production efficiency and quality.

➤ Flexible production

Intelligent manufacturing production lines adapt to diverse and personalized market demands, allowing for quickly switch production modes and adjust production plans. This enables small-batch, multi-variety production, reducing costs and enhancing competitiveness.

➤ Green and eco-friendly

Intelligent manufacturing production lines focus on energy-saving and emission reduction by using eco-friendly equipment and processes to lower energy consumption and pollution. Optimized processes and resource efficiency contribute to green manufacturing and a circular economy.

➤ Safety and reliability

Intelligent manufacturing production lines adopt advanced safety protection, encryption, and redundancy design to ensure stable operation and data security. Self-diagnosis and self-recovery features reduce the risk of production interruptions.

In the future, intelligent manufacturing production lines are expected to see widespread application and deeper derelopment worldwide, playing an increasingly important role in transforming and sustaining the manufacturing industry worldwide.

New Words and Phrases

arrangement[əˈreɪndʒmənt] *n.* 安排

utilize[ˈjuːtəlaɪz] *v.* 利用

optimize[ˈɒptɪmaɪz] *v.* 优化

digital[ˈdɪdʒɪtl] *adj.* 数字的

workshop[ˈwɜːrkʃɒp] *n.* 车间

encompass[ɪnˈkʌmpəs] *v.* 包含

integration[ˌɪntɪˈɡreɪʃ(ə)n] *n.* 整合

hardware[ˈhɑːdweə(r)] *n.* 硬件

software[ˈsɒftweə(r)] *n.* 软件

platform[ˈplætfɔːm] *n.* 平台

network[ˈnetwɜːk] *n.* 网络

visualization[ˌvɪzjuəlaɪˈzeɪʃn] *n.* 可视化

control[kənˈtroʊl] *v.* 控制

detection[dɪˈtɛkʃn] *v.* 检测

warehousing[ˈweəhaʊsɪŋ] *n.* 仓储

emission[ɪˈmɪʃn] *n.* 排放

reliability[rɪˌlaɪəˈbɪlɪti] *n.* 可靠性

encryption[ɪnˈkrɪpʃn] *n.* 加密

redundancy[rɪˈdʌndənsɪ] *n.* 冗余

be linked with　与……相联系

with the help of　在……的帮助下

adapt to　适应

focus on　专注于

playing an increasingly important role in　在……中扮演着越来越重要的角色

Technical Terms

intelligent manufacturing production line　智能制造生产线

intelligent management　智能管理

intelligent monitoring　智能监控

intelligent processing　智能加工

intelligent assembly　智能装配

intelligent inspection　智能检测

intelligent logistics　智能物流

the Industrial Internet　工业互联网

manufacturing execution systems (MES)　制造执行系统

enterprise resource planning (ERP)　企业资源规划

intelligent warehousing　智能仓储

Task 9 *Read the passage and match the two parts of the sentences.*

1. An intelligent manufacturing production line is	A. hardware devices, software systems, and data platforms.
2. An intelligent manufacturing production line covers	B. energy-saving and emission reduction.
3. An intelligent manufacturing production line encompasses	C. diverse and personalized market demands.
4. An intelligent manufacturing production line integrates	D. automated equipment, digital workshops, and intelligent factories.
5. An intelligent manufacturing production line sees	E. intelligent management, intelligent monitoring, intelligent processing, intelligent assembly, intelligent inspection, and intelligent logistics.
6. An intelligent manufacturing production line adapts to	F. advanced safety protection, encryption, and redundancy design to ensure stable operation and data security.
7. An intelligent manufacturing production line focuses on	G. a highly automated arrangement.
8. An intelligent manufacturing production line adopts	H. widespread widespread application and deeper derelopment worldwide.

Task 10 ***Translate the sentences that you completed in Task 9. The first one is provided.***

1. An intelligent manufacturing production line is a highly automated arrangement.
智能制造生产线是一种高度自动化的组织形式。

2.

3.

4.

5.

6.

7.

8.

Task 11 *Listen to a recording about industrial robots. Fill in the blanks with the words provided in the box.*

• automated	• versatility	• robotic	• warehouses
• medical	• revolutionized	• productive	• assemble

Industrial robots are 1. ________ machines used in intelligent manufacturing production line. These multipurpose machines typically consist of at least one reprogrammable 2. ________ arm, or a manipulator, that operates on three axes or more. Their ability to execute repetitive tasks at rapid speeds with high precision and without breaks has led to widespread implementation in 3. ________ and on factory assembly lines. Industrial robots manufacture and 4. ________ products, take on highly repetitive tasks, and even assist in groundbreaking 5. ________ procedures. This 6. ________ has made robotics prevalent across nearly every industry. Industrial robots have undeniably 7. ________ modern manufacturing, enabling safer, more 8. ________, and higher-quality production by taking on dull, dirty, and dangerous jobs that were previously done by manual workers.

Researching

▶ How are industrial robots applied in China's manufacturing industry? Let's study and analyze some cases together.

中国的工业机器人在制造业中有着怎样的应用？让我们一起研究和分析一些案例。

> 2021年12月，中国政府联合15家机关部门发布了《"十四五"机器人产业发展规划》(*the 14th Five-Year Plan for Robotics Industry*)，明确了机器人产业规划的重大意义并提出了机器人产业规划的目标，将中国机器人产业再一次推向新的高度。
>
> 国际机器人联合会(International Federation of Robotics)统计数据显示，2022年全球工业机器人新增装机量53.1万台，再创历史新高。我国工业机器人装机量超过全球总量的50%，连续9年位居世界首位。

Task 12 *Cases Study: The Main Application of Industrial Robots in Manufacturing in China (Fig. 1.2).*

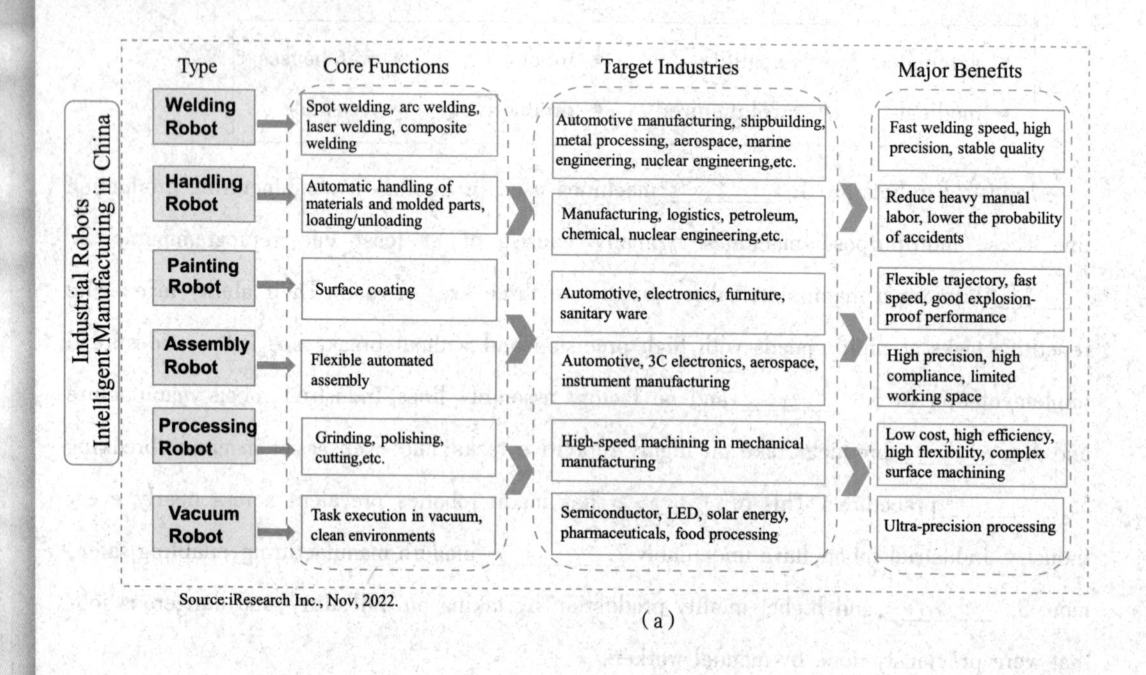

(a)

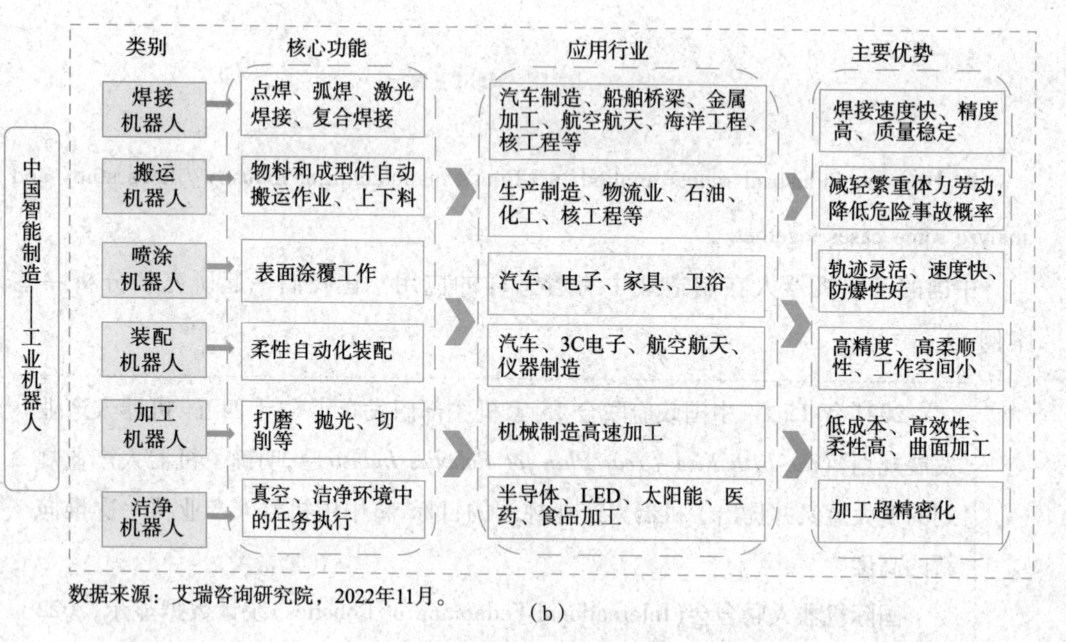

(b)

Fig. 1.2 The main application of industrial robots in manufacturing in China

(a) English version; (b) Chinese version

Task 13 ***Case Analysis: Analyze the main types of industrial robots produced by a particular Chinese Coupany, as well as their target industries and main advantages.***

Building and Presenting

Building

实施

▶ You need to create a poster promoting the use of industrial robots in intelligent manufacturing for ABC Company. As a group of 4 – 6 members, you may utilize AI technology, online research, fieldwork, and the knowledge gained from this unit to gain a deeper understanding of the topic. Then work through the following steps in Fig. 1.3 to help you complete your project.

制作一张海报,宣传 ABC 公司的工业机器人在智能制造中的应用。请以小组(4 ~6 名成员)为单位,通过使用 AI 技术、网络调研、实地考察等方法,结合本单元所学知识,深入了解主题,并按照图 1.3 中的步骤流程完成你的海报设计。

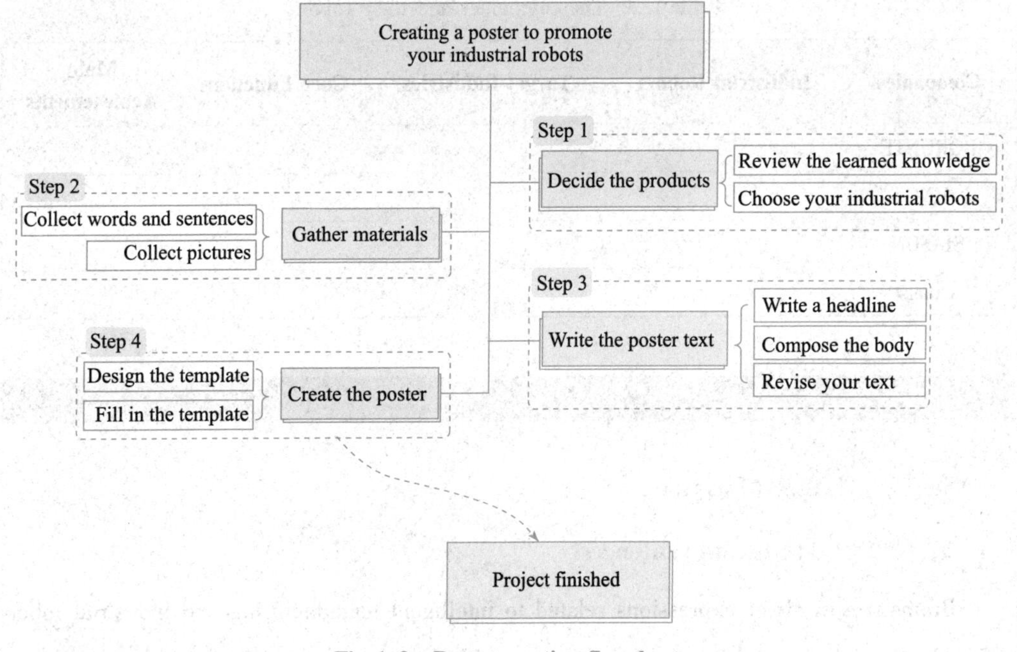

Fig. 1.3 Poster creation flowchart

Task 14

Step 1 Decide the Products

A. Review the learned knowledge

Answer the following questions in Table 1. 1 based on what you've learned before.

Table 1. 1 List of questions under two different concepts

Topics	Questions	Notes
Intelligent manufacturing production	What is intelligent manufacturing?	
	What are the key components of intelligent manufacturing production?	
	What is an intelligent manufacturing production line?	
Industrial robots in manufacturing	How many types of industrial robots do you know?	
	How are industrial robots used?	
	What are the advantages of industrial robots?	

B. Choose your industrial robots

Search for information about the following Chinese companies in Table 1. 2, select one of them, and complete the content in the table. Or you can pick another one as you like.

Table 1. 2 The companies and their robots

Companies	Industrial Robots	Target Industries	Core Functions	Main Achievements
BORUNTE				
HNC				
SIASUN				
Others				

Task 15

Step 2 Gather Materials

A. Collect words and sentences

Brainstorm words or expressions related to intelligent manufacturing and industrial robots (Figs. 1. 4 and 1. 5), and then make sentences with them.

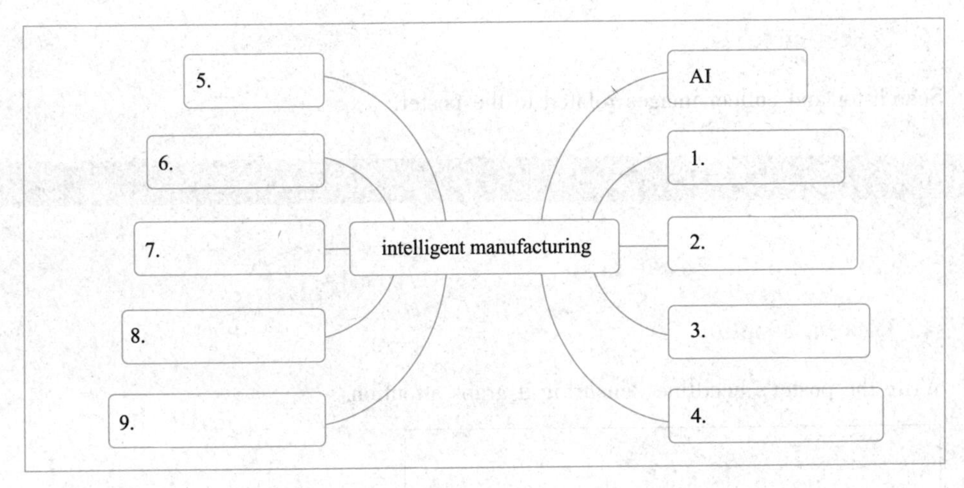

Fig. 1. 4 Words and expressions related to intelligent manufacturing

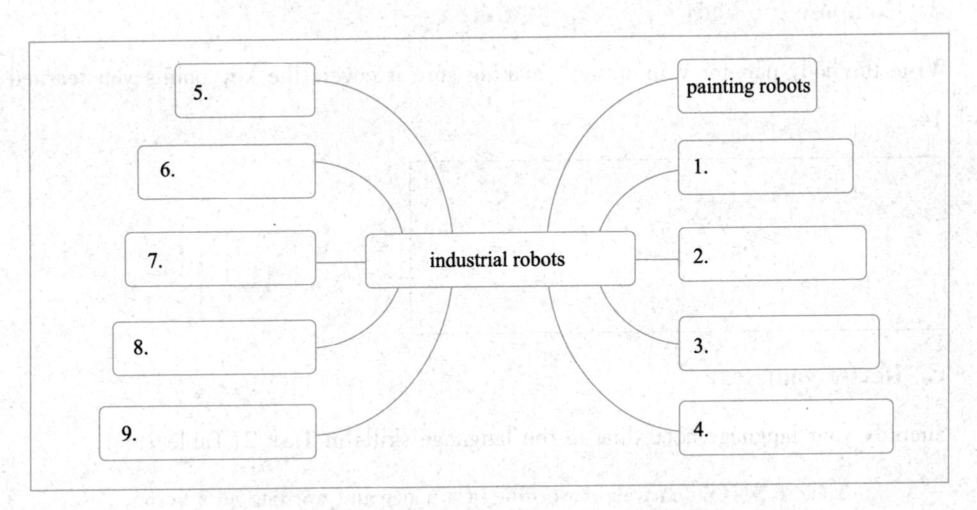

Fig. 1. 5 Words and expressions related to industrial robots

Example: BORUNTE painting robots are applied in different industries such as automobile, railway, building materials, home appliance, machinery, etc.

1.

2.

3.

4.

5.

6.

7.

B. Collect pictures

Search for and collect images related to the poster.

Task 16

Step 3 Write the Poster Text

A. Write a headline

Write the poster's headline, ensuring it grabs attention.

B. Compose the body

Write the body part for your poster, making sure it covers the key points you learned in Task 14.

C. Revise your text

Simplify your language according to the language skills in Task 2 (Table 1.3).

Table 1.3 Comparison of wording in a paper and wording on a poster

Wording in a paper	Wording on a poster

Task 17

Step 4 Create the Poster

A. Design the template

Use AI to design a poster template. The template in Fig. 1.6 is for your reference only.

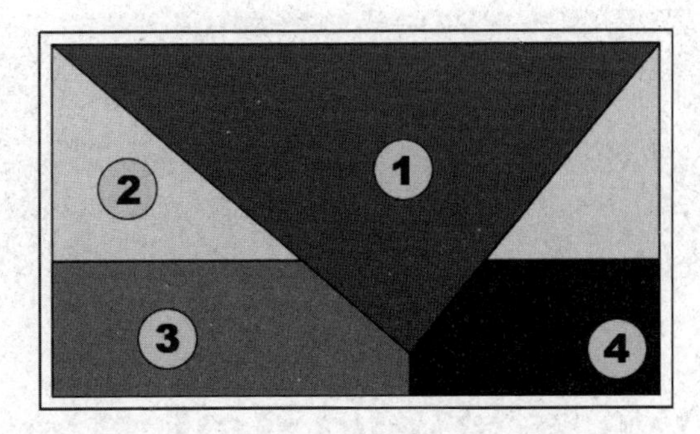

Fig. 1.6　Poster Layout design diagram

①Main focus area: Location of research fundamentals.

②Secondary emphasis: Location of important information.

③Supporting area: Location of supporting information.

④Final information area: Location of supplemental information.

B. Fill in the template

Place the text and pictures into the template you designed and finalize the poster.

Presenting
展示

▶ Now that you have finished your poster, it is time to share your work with your classmates.

海报完成后,请与同学分享你们的成果。

Task 18 *As a group, share with your class about your poster. According to the schedule in Table 1.4, take questions, comments, or suggestions from the audience if possible.*

Table 1.4　Task assignment

Leader	Student A	Student B	Student C	Student D
Opening and ending	Theme	Linguistic skills	Technical strategies	Artistic strategies

Task 19 *The group leader sends the modified posters to the teacher's mailbox or uploads them to the online platform as the usual performance evaluation document and for everyone to learn from online.*

Assessing and Reflecting

Assessing

评估

▶ Assessing the work you have accomplished allows you to know not only the response of your audience to your work but also the weak parts of your work to improve.

评估你所完成的工作，不仅能了解观众对你作品的反馈，还能发现不足，以便改进。

Task 20 *The group project is evaluated jointly by teachers and students(Table 1.5).*

Table 1.5 Group project evaluation form

Evaluation (*Excellent* =3, *Good* =2, *Poor* =1)					
Group	**Language**	**Creativity**	**Presentation**	**Collaboration**	**Cross-subject(poster)**
Group1					
Group2					
Group3					
Group4					
Group5					

Reflecting

反思

▶ Reflecting on your learning helps you understand what you've gained from this unit and how you can apply it in the future.

反思有助于你理解在本单元中所学的知识，以及将来如何加以运用。

Task 21 *The following questions in Table 1.6 may help you do the reflection, but feel free to ask yourself more questions when necessary.*

Table 1.6 Question reflection form

Questions	Reflections
Did I increase my vocabulary to finish my task more precisely?	
Did I use any digital technologies?	
Did I actively engage in preparing the project?	
Did I develop my presentation skills through presenting my group's ideas and work?	
Did my contributions help others understand production in IM better?	
Did I identify, analyze, and solve problems by fulfilling various tasks?	

Extending

▶ You are a front-line assembler at an automobile manufacturing plant. How would you operate a six-axis robotic arm to pick up and transport an auto part to the designated position on the assembly line?

你是某汽车生产厂的一线装配工人。你要如何操作六轴机械手臂来抓取零部件并将其运到装配线的指定位置呢?

Task 1 *Get acquainted with the robotic arms.*

In intelligent manufacturing, robotic arms are indispensable components of automated production lines. A six-axis robotic arm is an industrial robot with six degrees of freedom, capable of simulating the complex movements of a human arm. Fig. 1.7 is an example diagram of it.

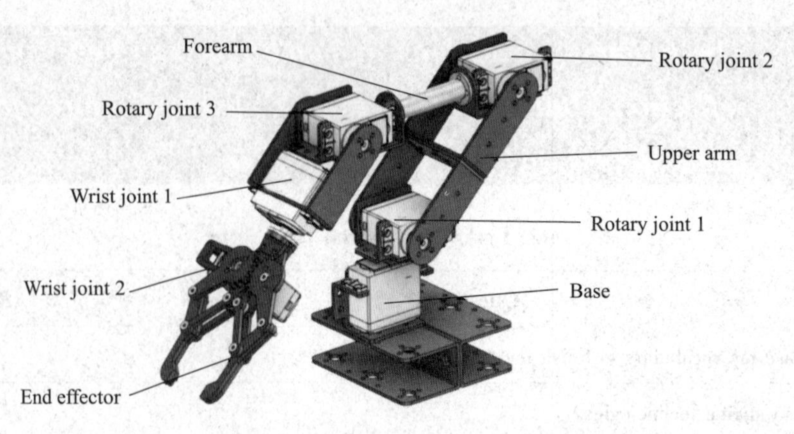

Fig. 1. 7 An industrial robotic arm

Task 2 *Familiarize yourself with the components and functions of a robotic arm (Table 1. 7).*

Table 1. 7 Robotic arm components and their functions

Name	Function
Base	The fixed part of the robotic arm, providing stability and support.
Rotary joint 1	Located above the base, enabling rotation on the base.
Upper arm	Connects Rotary joints 1 and 2, used for lifting and supporting the robotic arm.
Rotary joint 2	Located at the end of the upper arm, enabling rotation of the upper arm.
Forearm	Connects Rotary joints 2 and 3, allowing wrist rotation for precise operations.
Rotary joint 3	The elbow joint, enabling flexion and extension, allowing the forearm to rotate.
Wrist joint 1	The pitch joint of the wrist, connecting the forearm and the end effector, allowing up-and-down tilting.
Wrist joint 2	The yaw joint of the wrist, usually located at the end of the arm, allowing rotation of the end effector.
End effector	The tool at the end of the robotic arm, used for grasping or manipulating objects, such as grippers, suction cups, or welding guns.

Task 3 *Complete the operation process in Fig. 1. 8 based on Task 1 and Task 2.*

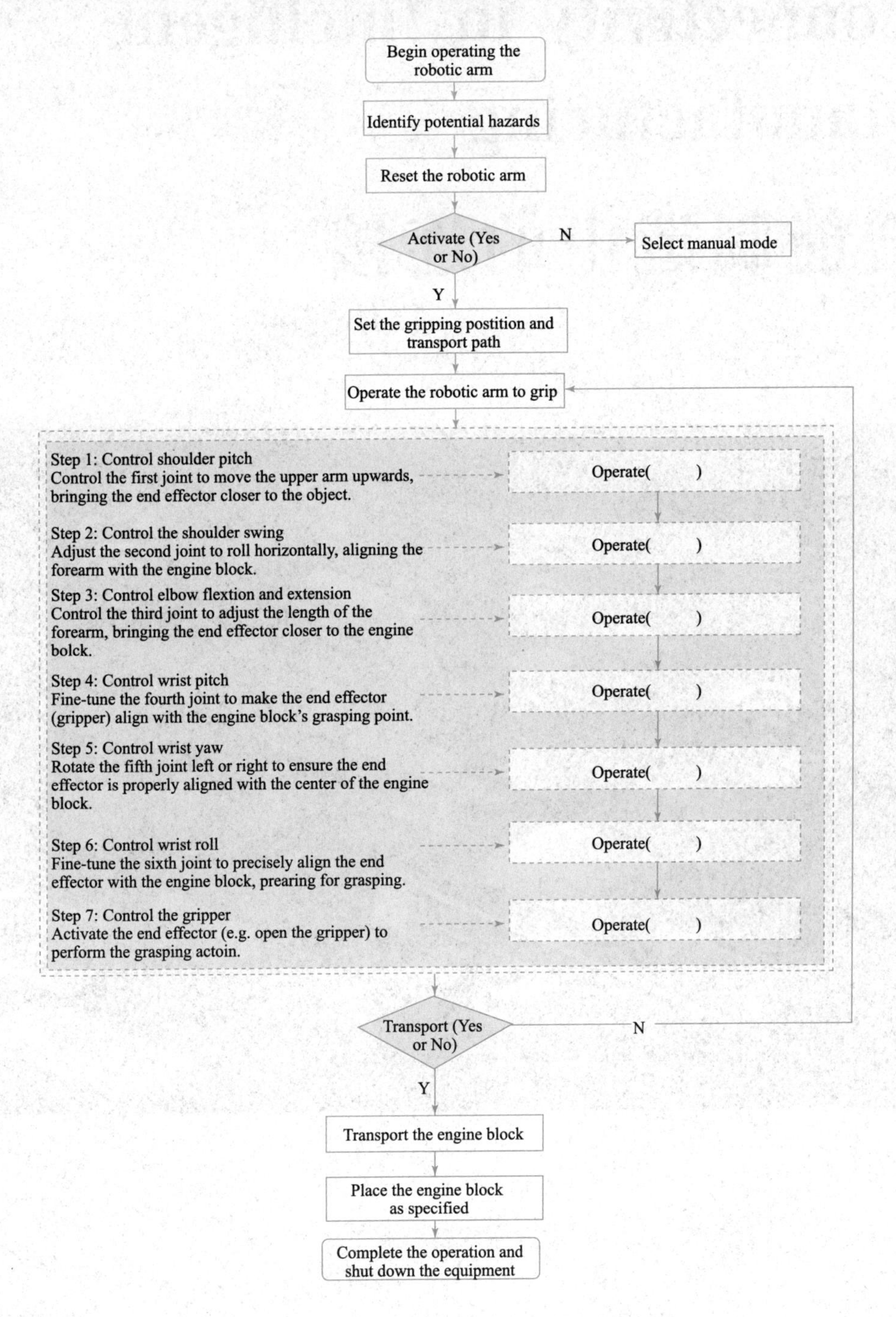

Fig. 1. 8 Robotic arm operation flowchart

Unit 2
Connectivity in Intelligent Manufacturing
智能制造中的连接

Unit Introduction

Driving Question

驱动问题

Flexible, reliable, and secure connectivity with low latency are critical components of intelligent manufacturing. By leveraging interconnected technologies, manufacturing enterprises can significantly enhance production efficiency and product quality. High-end CNC (computer numerical control) machines, as a prime example of intelligent manufacturing's inter connected applications, have greatly benefited from these interconnected technologies. How do these advancements translate into practical advantages for manufacturing?

灵活、可靠、安全、低延时的连接对智能制造至关重要。通过互联技术，制造企业能够显著提升生产效率和产品质量。高端数控机床作为智能制造互联应用中的典型案例，极大地受益于这些互联技术。那么，这些进步如何为制造业带来实际效益呢?

Project Overview

项目概述

In this unit, you will begin by exploring the concept of connected manufacturing. After that, you will delve into the role of inter connected platforms, such as the Industrial Internet of Things (IIoT). Subsequently, you will learn about the concept of CNC machines. Then, you will study practical application cases of China's high-end CNC machines and how they have benefited from these interconnected technologies. Next, working in groups, you will create a promotional video for ABC Company's high-end interconnected CNC machines. After the presentation, you will discuss the benefits and challenges of integrating inter connected technologies into manufacturing.

在本单元，首先，学习互联制造的概念。然后，深入了解互联平台(如工业物联网)的作用并学习数控机床的概念。之后，研究中国高端互联数控机床的实际应用案例，了解它们是如何通过互联技术提高性能的。随后，小组合作为 ABC 公司的高端数控机床制作一段宣传视频并进行展播。最后，讨论将互联技术应用到制造业中的优点和挑战。

Project Extension

项目拓展

After completing the main project, you'll engage in a hands-on extension activity to deepen your understanding of high-end interconnected CNC machines. This will involve learning how to operate a CNC machine to perform simple tasks.

完成主要项目后，将进行实操拓展，学习操作机床并完成简单的任务，以进一步巩固对高端互联数控机床的理解。

Learning Objectives

学习目标

This unit is intended to help you:

1. have a general idea of connected manufacturing;
2. explore the key technologies and benefits of IIoT;
3. analyze cases of high-end CNC machines in simple English;
4. create and present an English video for high-end interconnected CNC machines;
5. operate a CNC machine;
6. develop a sense of security regarding networking.

本单元旨在帮助你：

1. 了解互联制造；
2. 探讨工业物联网的技术和益处；
3. 用英语简单分析中国高端数控机床的应用案例；
4. 制作并展示高端互联数控机床的英文视频；
5. 操作数控机床；
6. 培养联网安全意识。

Getting Ready

► What is a promotional video, and what basic linguistic features define its script? Please complete the tasks below to find the answer. Fig. 2. 1 is screenshot of a video playing.

什么是宣传视频？其脚本在语言方面有哪些基本特点？请完成以下任务，找到答案。图 2. 1 是一张视频截屏图片。

Fig. 2.1 Screenshot of a video playing

Task 1 *Listen to the passage about a promotional video and complete the sentences with the words you hear.*

1. A promotional video is a short video, usually between ________________.

2. If you're aiming for ________________, you should consider diving into promotional videos rather than some quick text or images.

3. Make sure they ________________ and spark their interest in the first few seconds.

4. Keep the content tightly focused on your ________________.

5. Choose background music that complements the video's mood without overshadowing the ________________.

Task 2 *Learn some linguistic features of a promotional video script.*

2.1 主动语态(Active Voice)

特点:主动语态突出执行者,使句子更加简洁直接,有助于提升宣传视频的感染力和节奏感。

示例: The factory produces ten new models each day.

2.2 被动语态(Passive Voice)

特点:被动语态强调结果,使宣传内容更具权威性和说服力。

示例: Ten new models are produced each day.

Task 3 *Rewrite the following sentences according to the requirements.*

1. Original: The team designs innovative packaging solutions.
   Passive Voice: ______________________________.
2. Original: The factory produces fifty units per hour.
   Passive Voice: ______________________________.
3. Original: Our company develops cutting-edge automation systems.
   Passive Voice: ______________________________.
4. Original: Engineers test all products rigorously before release.
   Passive Voice: ______________________________.
5. Original: The project manager oversees the construction of the new plant.
   Passive Voice: ______________________________.
6. Original: Technicians install the latest software updates on machinery.
   Passive Voice: ______________________________.
7. Original: New safety protocols are implemented by the safety officer.
   Active Voice: ______________________________.
8. Original: The new product line is managed by a team of experts.
   Active Voice: ______________________________.
9. Original: Quality control is maintained by continuous monitoring.
   Active Voice: ______________________________.

10. Original: The final assembly is completed by skilled workers.

Active Voice: ______________________________.

11. Original: The blueprint is approved by the engineering department.

Active Voice: ______________________________.

12. Original: The shipment of materials is coordinated by the logistics team.

Active Voice: ______________________________.

Starting up

▶ Connected manufacturing is becoming a central focus in the manufacturing industry. What have you already known about it? In this section, we'll learn about three types of connected manufacturing.

互联制造正逐渐成为制造业的焦点。你对它了解多少？在本部分，我们将学习互联制造的三种类型。

Connected Manufacturing

Connected manufacturing, unlike traditional manufacturing, leverages advanced technologies such as the Internet of Things (IoT), AI, and cloud computing. This integration enables intelligent communication, automation, and control of manufacturing processes, providing manufacturers with comprehensive visibility into their people, machines, and processes. And this visibility helps identify what's working and what isn't, enabling critical changes at every organizational level.

Creating a connected manufacturing environment involves implementing technologies from three distinct categories: machine-to-machine (M2M), machine-to-human (M2H), and human-to-human (H2H) communication.

➢ Machine-to-machine communication

M2M communication is essential for smart manufacturing, allowing devices and systems to exchange information and act autonomously without human intervention. This communication enables machines to "talk" to each other on the shop floor, often relying on original equipment manufacturer (OEM) embedded devices within machines and equipment that communicate with Industrial Internet of Things (IIoT) platforms. Programmable Logic Controllers (PLCs) and edge computing devices are also commonly used in M2M communication.

➢ Machine-to-human communication

M2H communication facilitates interaction between workers and machines, ensuring operators, technicians, and managers are informed about machine status, performance metrics, alerts, and other relevant data. Technologies like human-machine interfaces (HMIs), mobile devices, and Augmented Reality (AR)/Virtual Reality (VR)-enabled devices make this communication possible. HMIs are touchpoints used in human-machine interactions to display machine performance data, with common applications including dashboards, push-button replacement screens, and system overview screens.

➢ Human-to-human communication

H2H communication is crucial for efficient coordination, collaboration, and decision-making within the connected manufacturing environment. There's no substitute for direct communication with coworkers on the shop floor. Communication and collaboration tools are core technologies of connected manufacturing, designed to streamline H2H and M2H communication using advanced technologies like data analysis, cloud computing, and IIoT networks. These tools empower front-line workers with real-time information that's easily accessible when and where they need it.

To sum up, utilizing technologies such as IoT, AI, and machine learning is a powerful way for manufacturers to improve efficiency, reduce downtime, control quality, increase safety, reduce costs, and achieve greater flexibility. As manufacturing becomes more competitive, connected manufacturing will play a critical role in ensuring success.

New Words and Phrases

enhance[ɪnˈhæns] *v.* 提高

leverage[ˈlevərɪdʒ] *v.* 利用

identify[aɪˈdentɪfaɪ] *v.* 识别

enable[ɪˈneɪbl] *v.* 使能够

involve[ɪnˈvɒlv] *v.* 涉及

implement[ˈɪmplɪment] *v.* 实施

distinct[dɪˈstɪŋkt] *adj.* 清楚的

autonomously[ɔːˈtɒnəməslɪ] *adv.* 自主地

intervention[ˌɪntəˈvenʃ(ə)n] *n.* 干涉

facilitate[fəˈsɪlɪteɪt] *v.* 促进

interaction[ˌɪntər'ækʃ(ə)n] n. 交互

inform[ɪn'fɔːrm] v. 通知

alert[ə'lɜːt] n. 警报

coordination[kəʊˌɔːdɪ'neɪʃn] n. 协调,配合

collaboration[kəˌlæbə'reɪʃ(ə)n] n. 合作,协作

streamline['striːmlaɪn] v. 简化,精简

empower[ɪm'paʊə(r)] v. 授权,准许

downtime['daʊntaɪm] n. 停工期,停机时间

rely on　依靠

make... possible　让……成为可能

be crucial for　至关重要

there's no substitute for　无可替代

to sum up　总的来说

Technical Terms

Internet of Things(IoT)　物联网

Industrial Internet of Things(IIoT)　工业物联网

cloud computing　云计算

machine-to-machine(M2M)　机器对机器

original equipment manufacturer(OEM)　原始设备制造商

machine-to-human(M2H)　机器对人

human-to-human(H2H)　人对人

Programmable Logic Controllers(PLCs)　可编程逻辑控制器

edge computing　边缘计算

human-machine interfaces(HMIs)　人机界面

Task 4　*According to the passage, are the following statements true (T) or false (F)?*

(　　) 1. There is no difference between traditional manufacturing and connected manufacturing.

(　　) 2. Connected manufacturing uses IoT, AI, and cloud computing to improve the performance of a factory.

(　　)3. Connected manufacturing provides manufacturers with limited visibility into their people, machines, and processes.

(　　)4. M2M communication enables devices and systems to exchange information and act autonomously without human intervention.

(　　)5. HMIs are not typically used in M2H communication to facilitate the display of machine performance data.

(　　)6. PLCs and edge computing devices are commonly used in M2M communication.

(　　)7. H2H communication is not essential in the connected manufacturing environment.

(　　)8. IoT, AI and machine learning technologies can help manufacturers improve efficiency, control quality and reduce costs.

Task 5 *Read the passage and answer the following questions.*

1. What is the main advantage of connected manufacturing over traditional manufacturing?

____.

2. What technologies are involved in creating a connected manufacturing environment?

____.

3. What is M2M communication?

____.

4. How does M2H communication become possible?

____.

5. Why is H2H communication important in connected manufacturing?

____.

Task 6 *Match the English expressions in Column A with their Chinese meanings in Column B.*

Column A	Column B
1. connected manufacturing	A. 云计算
2. cloud computing	B. 人机界面
3. Internet of Things (IoT)	C. 虚拟现实
4. M2H communication	D. 机器对人通信
5. Virtual Reality (VR)	E. 互联制造
6. Programmable Logic Controllers (PLCs)	F. 可编程逻辑控制器
7. human-machine interfaces (HMIs)	G. 边缘计算
8. edge computing	H. 物联网

Task 7 *Make six sentences with words from the following boxes.*

enhance   leverage   identify   implement

involve   empower   inform   streamline

Example: The use of advanced technology can **enhance** the efficiency of manufacturing processes.

1. ______________________________.
2. ______________________________.
3. ______________________________.
4. ______________________________.
5. ______________________________.
6. ______________________________.

Task 8 *Translate the following sentences into English using the words and expressions given.*

1. 实现可持续发展有赖于各级部门的共同努力。(rely on)

____.

2. 对于应届生而言,提升自我对于找工作尤为重要。(be crucial for)

____.

3. 科技的进步使得中产阶级能够享受以前只有富人才有能力负担的东西。(make sth. possible)

____.

4. 机器与人类的协作提高了生产效率和产品质量。(collaboration)

____.

5. 我相信科技能为残障人士赋能。(empower)

____.

Investigating

▶ Central to connected manufacturing is the Industrial Internet of Things (IIoT), which connects machines, devices, and systems. In this section, we will explore how IIoT supports manufacturing.

实现制造的互联需要工业物联网,它能连接机器、设备和系统。本部分,我们将探讨工业物联网是如何支持制造的。

The Industrial Internet of Things

IIoT is a specialized subset of IoT, specifically designed for industrial applications. As industries increasingly merge information technology (IT) with operational technology (OT), IIoT has become a robust network linking equipment and devices via sensor technology. At its core, IIoT employs smart sensors, actuators, and devices like Radio Frequency Identification (RFID) tags to enhance industrial and manufacturing processes.

Key Technologies in IIoT Architecture

➢ Sensors and devices

Sensors and devices are the cornerstone of IIoT, capturing data from the physical environment and converting it into digital format. They monitor a variety of parameters, such as temperature, pressure, and motion, and transmit this data to central systems for analysis.

➢ Connectivity technologies

Connectivity technologies are crucial for data transfer in IIoT systems, employing technologies like Wi-Fi, cellular networks, satellite communication, and Low-power Wide-area Networks (LPWANs). The choice of technology depends on factors like data volume, range, and power consumption.

➢ Messaging Protocols

Messaging Protocols establish the communication language within IIoT systems, ensuring effective data exchange. Protocols like Message Queuing Telemetry Transport (MQTT) are favored for their reliability and efficiency in various network conditions.

➢ Edge computing

Edge computing tackles data volume and latency issues by processing data near its source, reducing transmission needs and improving response times. This is especially advantageous in real-time industrial applications.

➢ Cloud platforms

Cloud platforms offer scalable and flexible infrastructure for data storage, processing, and analysis. They facilitate advanced data analysis, machine learning, and AI, unlocking valuable insights from extensive data sets.

➢ Data analysis and AI

Data analysis and AI are pivotal in IIoT, transforming raw data into actionable insights. These technologies enable predictive maintenance, process optimization, and cost reduction, significantly enhancing industrial efficiency and decision-making.

Key Benefits of IIoT

IIoT's networked sensors and actuators allow companies to detect inefficiencies and issues early, saving time and money while enhancing business intelligence efforts. In manufacturing, IIoT aids in quality control, sustainable practices, supply chain traceability, and overall efficiency. It plays a vital role in processes like predictive maintenance, enhanced field service, energy management, and asset tracking, ensuring a holistic approach to industrial management.

In conclusion, IIoT represents a revolutionary approach to industrial operations, leveraging advanced technologies to improve efficiency, reliability, and economic growth.

New Words and Phrases

application[ˌæplɪˈkeɪʃ(ə)n] *n.* 运用

device[dɪˈvaɪs] *n.* 设备

employ[ɪmˈplɔɪ] *v.* 使用

actuator[ˈæktʃʊeɪtər] *n.* 执行器

cornerstone[ˈkɔːnəstəʊn] *n.* 基石

capture[ˈkæptʃə(r)] *v.* 捕获

convert[kənˈvɜːt] *v.* 转换

monitor[ˈmɒnɪtə(r)] *v.* 监测

parameter[pəˈræmɪtə(r)] *n.* 参数

motion[ˈməʊʃn] *n.* 运动

transmit[trænzˈmɪt] *v.* 传输

protocol[ˈprəʊtəkɒl] *n.* 协议

tackle[ˈtæk(ə)l] *v.* 解决

latency[ˈleɪtənsi] *n.* 延迟

scalable[ˈskeɪləbl] *adj.* 可扩展的

infrastructure[ˈɪnfrəstrʌktʃər] *n.* 基础设施

unlock[ʌnˈlɒk] *v.* 解锁

pivotal[ˈpɪvət(ə)l] *adj.* 关键的

actionable[ˈækʃənəbl] *adj.* 可操作的

holistic[hoʊˈlɪstɪk] *adj.* 全面的

revolutionary[revəˈluːʃənərɪ] *adj.* 革命性的

merge with　与……融合

transfer. . . to　转换为

transform. . . into　转变成

Technical Terms

information technology(IT)　信息技术

operational technology(OT)　运营技术

radio frequency identification(RFID)　射频识别

cellular networks　蜂窝网络

Low-power Wide-area Networks(LPWANs)　低功耗广域网

Messaging Protocols　消息协议

Message Queuing Telemetry Transport(MQTT)　消息队列遥测传输

Task 9　*Read the passage and match the two parts of the sentences.*

1. IIoT is	A. quality control, sustainable practices, supply chain traceability, and overall efficiency.
2. IIoT has become	B. the communication language within IIoT systems, ensuring effective data exchange.
3. IIoT aids in	C. raw data into actionable insights.
4. Sensors and devices capture	D. scalable and flexible infrastructure for data storage, processing, and analysis.
5. Messaging Protocols establish	E. a specialized subset of IoT, specifically designed for industrial applications.
6. Edge computing tackles	F. data from the physical environment and converting it into digital format.
7. Cloud platforms offer	G. a robust network linking equipment and devices via sensor technology.
8. Data analytics and AI transform	H. data volume and latency issues by processing data near its source.

Task 10　*Translate the sentences that you completed in Task 9. The first one is provided.*

1. IIoT is a specialized subset of IoT, specifically designed for industrial applications.
工业物联网是物联网的一个专门子集,专为工业应用而设计。

2.

3.
4.
5.
6.
7.
8.

Task 11 ***Listen to a recording about CNC machines. Fill in the blanks with the words provided in the box and read it aloud.***

• material	• manual	• software	• control
• crucial	• movements	• connected	• operation

CNC stands for computer numerical 1. ________, which refers to the computerized 2. ________ of machining tools used in manufacturing. CNC machines operate using pre-programmed 3. ________ and codes, which tell each machine the exact 4. ________ and tasks to complete. For example, a CNC machine might cut a piece of 5. ________ (such as metal or plastic) based on instructions from a computer, meeting the specifications pre-coded into the program—all without a 6. ________ machine operator. Once 7. ________, CNC machines can enhance visibility into manufacturing efficiency and allow for remote data access. This connectivity is 8. ________ for improving productivity and reducing downtime.

Researching

▶ How are connected high-end CNC machines being applied in China? Let's study and analyze some cases together.

在中国,高端互联数控机床是如何应用的?让我们一起研究和分析一些案例。

数控机床被称为"工业之母"。数控机床行业在国家政策的支持和企业的不断创新下,呈现出快速发展的态势。各主要数控机床企业,包括创世纪、华中数控、秦川机床、海天精工、宇环数控等,均在加快高端数控机床的布局。国产数控系统企业已占领了我国经济型数控系统95%以上的市场份额,为我国的数控系统产业发展做出了巨大贡献。

Task 12 *Case Study: The Application of Connected High-end CNC Machines in Manufacturing in China (Fig. 2. 2).*

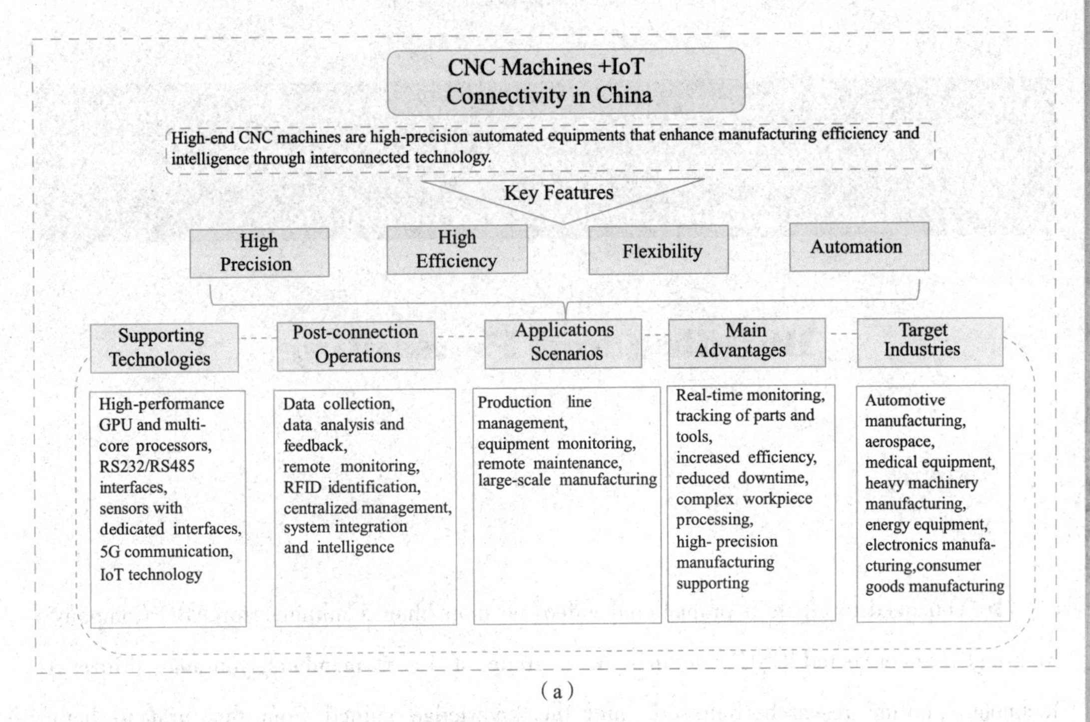

(a)

Fig. 2. 2 The application of connected high-end CNC machines in manufacturing in China

(a) English version

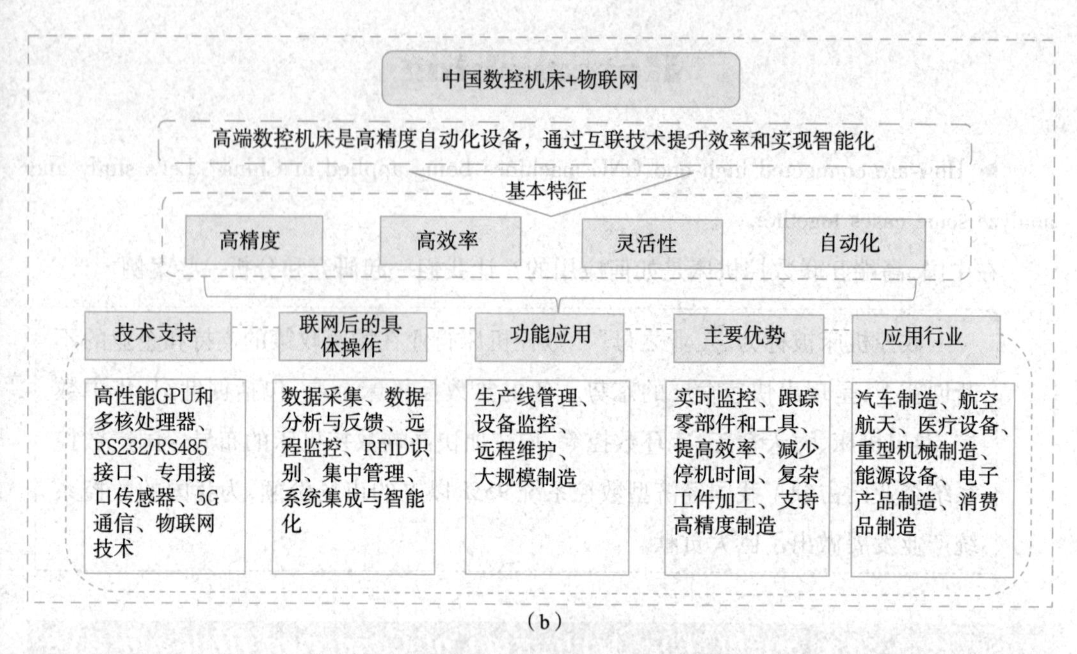

(b)

Fig. 2.2 The application of connected high-end CNC machines in manufacturing in China(Continued)

(b) Chinese version

Task 13 ***Cases Analysis: Analyze the main types of CNC machines produced by Chinese companies like QCMT&T, as well as their target industries and main advantages.***

Building and Presenting

Building
实施

▶ You need to create a promotional video (no more than 5 minutes) for ABC Company's high-end interconnected CNC machines. As a group of 4 – 6 members, you may utilize AI technology, online research, fieldwork, and the knowledge gained from this unit to better understand the topic. Then work through the following steps in Fig. 2.3 to help you complete the project.

制作一段视频,宣传 ABC 公司的高端互联数控机床(时长不超过 5 分钟)。请以小组(4~6 名成员)为单位,通过使用 AI 技术、网络调研、实地考察等方法,结合本单元所学知识,深入了解主题,并按照图 2.3 中的步骤流程完成你的宣传视频。

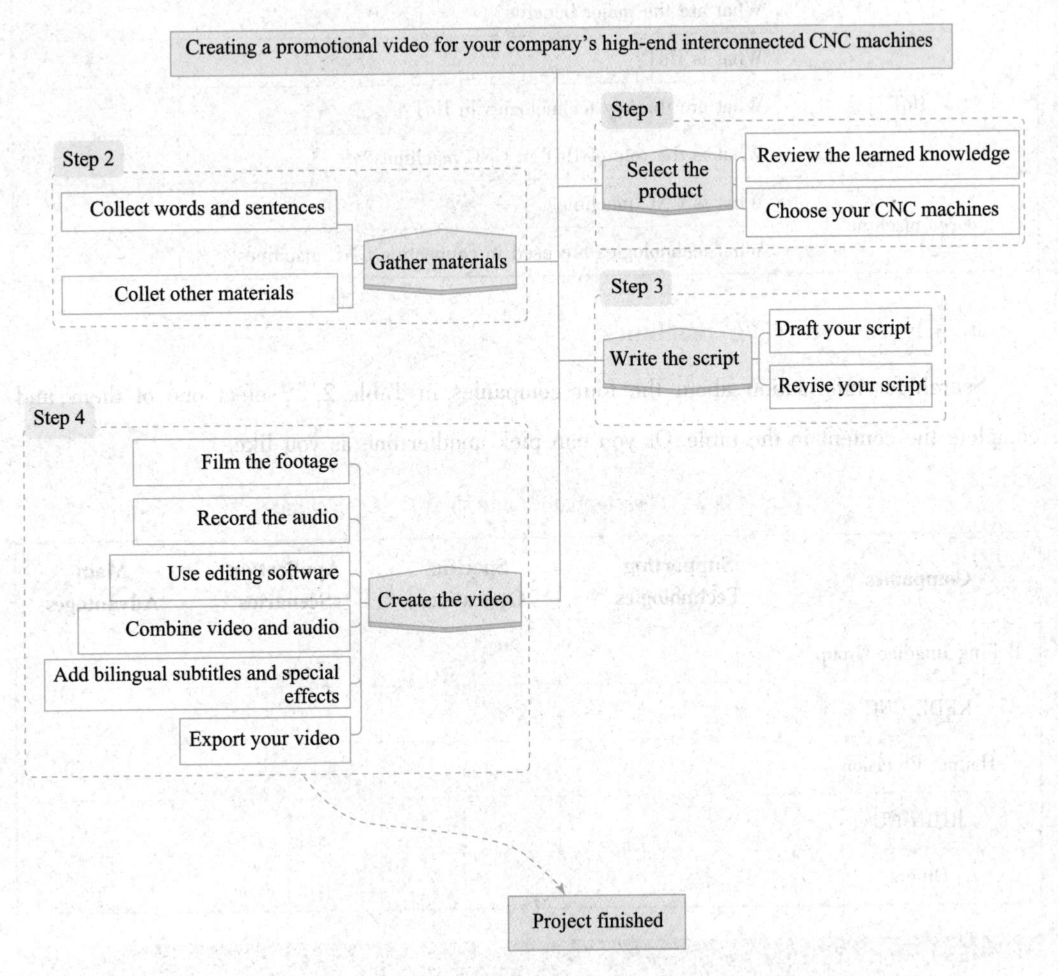

Fig. 2.3 Video creation flowchart

Task 14

Step 1 Select the Product

A. Review the learned knowledge

Answer the following questions in Table 2.1 based on what you've learned before.

Table 2.1 List of questions under three different topics

Topics	Questions	Notes
Connected manufacturing	What is connected manufacturing?	
	What are the major benefits?	
IIoT	What is IIoT?	
	What are the key technologies in IIoT?	
	What is the role of IIoT in CNC machines?	
CNC machine	What is CNC machine?	
	What technologies are used in connecting CNC machines?	

B. Choose your CNC machines

Search for information about the four companies in Table 2.2, select one of them, and complete the content in the table. Or you can pick another one as you like.

Table 2.2 The companies and their CNC machines

Companies	Supporting Technologies	Specific Operations	Application Scenarios	Main Advantages
Beijing Jingdiao Group				
KEDE CNC				
Haitian Precision				
RIJIN-NC				
Others				

Task 15

Step 2 Gather Materials

A. Collect words and sentences

Brainstorm words or expressions related to connected manufacturing and CNC machine (Figs. 2.4 and 2.5), and then make sentences with them.

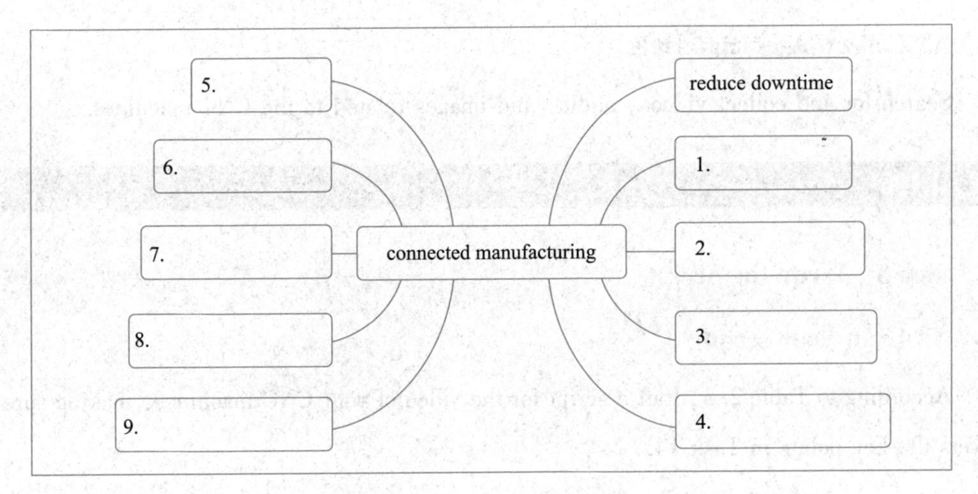

Fig. 2.4 Words and expressions related to connected manufacturing

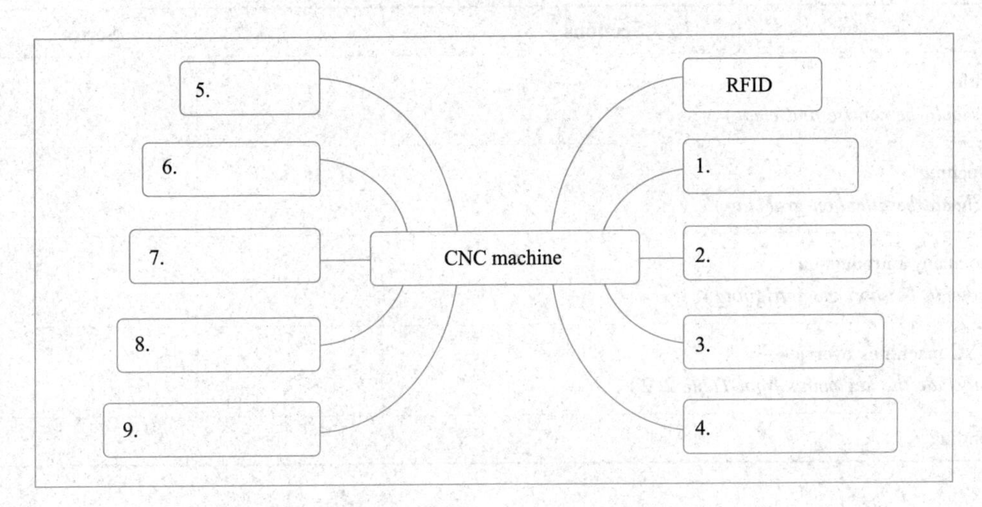

Fig. 2.5 Words and expressions related to CNC machine

Example: Zhejiang RIFA Digital Precision Machinery Co., Ltd. specializes in research and development, manufacturing, as well as sales of CNC machines.

1.

2.

3.

4.

5.

6.

7.

B. Collect other materials

Search for and collect videos, audio, and images related to the CNC machines.

Task 16

Step 3 Write the Script

A. Draft your script

According to Table 2. 3, draft a script for the video of your CNC machines, making sure it covers the key points in Task 14.

Table 2. 3 Sections of a script and writing requirements

Sections	Script
Title (*should be concise and clear*)	
Opening (*should be attention-grabbing*)	
Company introduction (*should be short but intriguing*)	
CNC machines overview (*include the key points from Table* 2. 2)	
Ending	

B. Revise your script

Polish your script according to the language skills you have learned in Task 2 (Table 2. 4).

Table 2. 4 Comparison of wording in the script and wording after polishing

Wording in the script	Wording after polishing

Task 17

Step 4 Create the Video

After finishing your script, follow the steps in Table 2.5, and mark each one as accomplished as you go until you export your own video.

Table 2.5 Video production steps

Video Production Process	
• Write the script	(√)
• Film the footage	( )
• Record the audio	( )
• Use editing software	( )
• Combine video and audio	( )
• Add bilingual subtitles and special effects	( )
• Export your video	( )

Presenting

展示

▶ Now that you have made your own video, it is time to share your work with your classmates.

视频制作完成后,请与同学分享你们的成果。

Task 18 *As a group, play your video in class. According to the schedule in Table 2.6, take questions, comments, or suggestions from the audience if possible.*

Table 2.6 Task assignment

Leader	Student A	Student B	Student C	Student D
Opening and ending	Theme	Linguistic skills	Technical strategies	Artistic strategies

Task 19 ***The group leader sends the modified videos to the teacher's mailbox or uploads them to the online platform as the usual performance evaluation document and for everyone to learn from online.***

Assessing and Reflecting

Assessing

评估

▶ Assessing the work you have accomplished allows you to know not only the response of your audience to your work but also the weak parts of your work to improve.

评估你所完成的工作，不仅能了解观众对你作品的反馈，还能发现不足，以便改进。

Task 20 ***The group project is evaluated jointly by teachers and students(Table 2. 7).***

Table 2. 7 Group project evaluation form

Evaluation (*Excellent* = 3, *Good* = 2, *Poor* = 1)					
Group	**Language**	**Creativity**	**Presentation**	**Collaboration**	**Cross-subject(video)**
Group1					
Group2					
Group3					
Group4					
Group5					

Reflecting

反思

▶ Reflecting on your learning helps you understand what you've gained from this unit and how you can apply it in the future.

反思有助于你理解在本单元所学的知识，以及将来如何加以运用。

Task 21 *The following questions in Table 2.8 may help you do the reflection, but feel free to ask yourself more questions when necessary.*

Table 2.8 Questions reflection form

Questions	Reflections
Did I increase my vocabulary to finish my task more precisely?	
Did I use any digital technologies?	
Did I actively engage in preparing the project?	
Did I develop my presentation skills through presenting my group's ideas and work?	
Did my contributions help others understand connectivity in IM better?	
Did I identify, analyze and solve problems by fulfilling various tasks?	

Extending

▶ You are a front-line operator at an electronics manufacturing plant. Now, you need to produce a batch of × × model phone cases made of metal for the × × brand. How would you complete the production process of this batch of phone cases according to the provided parts drawings?

你是某电子加工厂的一线操作员，现在需要生产一批××品牌××型号的金属手机壳。你要如何根据提供的手机壳零件图纸完成该批次手机壳的生产和制作呢？

Task 1 *Get acquainted with the CNC machine center.*

This is a semi-closed loop vertical machining center with servo control for the X, Y, and Z axes. They are designed with a compact and well-proportioned structure, with the spindle driven by a servo motor. It is capable of performing multiple operations including drilling, milling, boring, reaming, and tapping on complex parts like disks, plates, housings, and molds in a single clamping. This machine is suitable for component processing in industries such as automotive parts, electrical instruments, motorcycles, tools, hardware, and motors. It can be used for small to medium-batch production and can also be integrated into an automated production line for mass production, ensuring product quality and production efficiency. Fig. 2. 6 shows an example of a CNC vertical machining center.

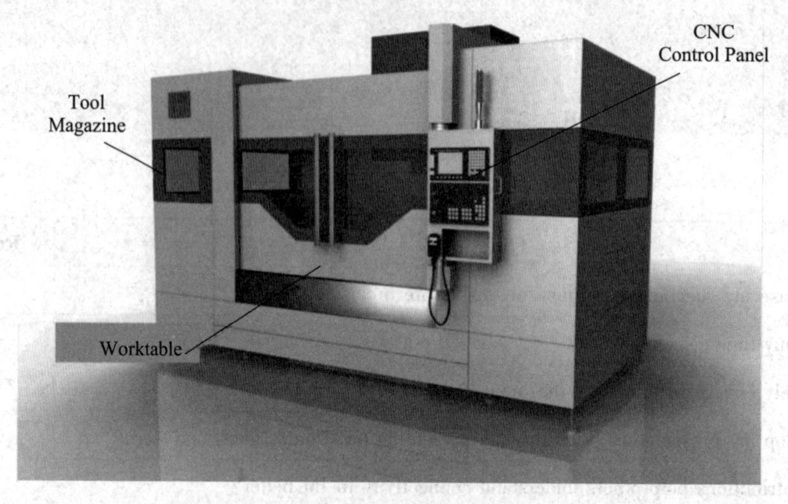

Fig. 2. 6 CNC vertical machining center

Task 2 *Familiarize yourself with the stages and procedures of the CNC machine operation.*

2. 1 Actual Operational Process of A CNC Machine (Fig. 2. 7).

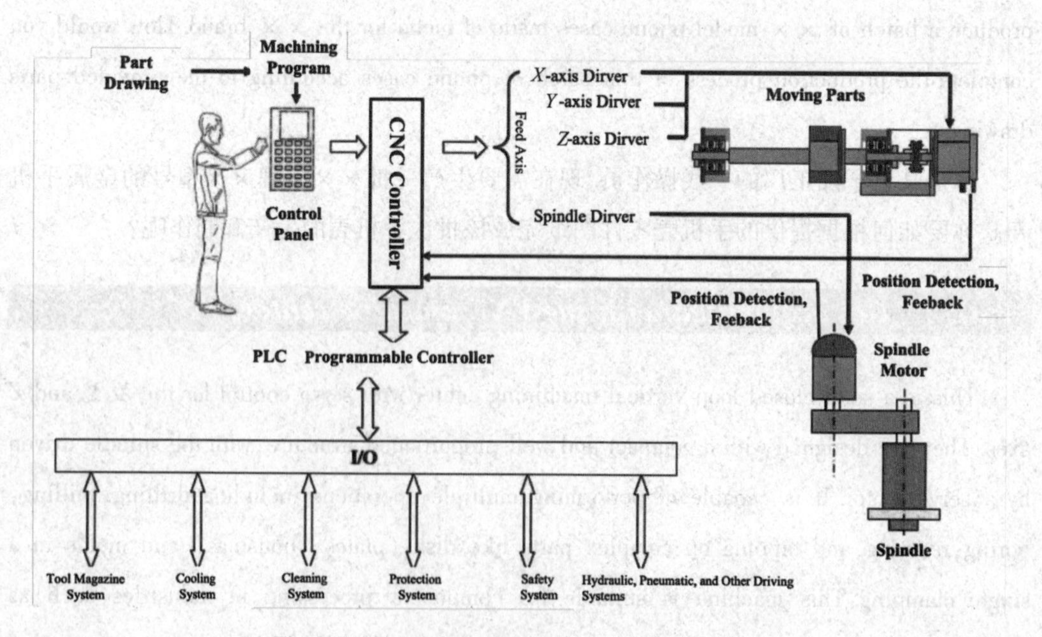

Fig. 2. 7 Operational process of a CNC machine

2.2 Operational Workflow for Manufacturing Phone Cases with A CNC Machine (Table 2.9)

Table 2.9 Operational workflow for manufacturing phone cases with A CNC machine

Operational Stages	Workflow
Design stage	Part drawing design: Define the dimensions, shape, material, and machining accuracy requirements for the phone cases.
Preparation stage	• Material preparation: Select the appropriate material according to the design requirements. • Programming and simulation: Use CAM (computer-aided manufacturing) software to develop the machining program based on the part drawing, and perform simulation verification. • Machine preparation: Select the appropriate CNC lathe and carry out machine debugging, including checking whether all parts of the machine are functioning properly, setting the tool change point, and adjusting the coordinate system.
Machining stage	• Workpiece clamping: Secure the phone cases blank or semi-finished product onto the CNC lathe fixture, ensuring accurate and stable positioning. • Tool selection and installation: Choose and install the appropriate tools. • Start machining: Proceed with rough machining, semi-finishing, and finishing. • Inspection and adjustment: Use measuring tools to check the machining progress and workpiece quality, and adjust machining parameters or tool positions as necessary.
Post-processing stage	• Cleaning and deburring: After machining is complete, clean the phone cases to remove surface oil and chips, and perform deburring to ensure a smooth and flawless surface. • Quality inspection: Conduct a comprehensive quality inspection of the finished phone cases, including dimensional accuracy, shape accuracy, surface quality, and functional testing. • Packaging and storage: Package the qualified phone cases, label them, and store them in the warehouse.

Task 3 *Complete the operation process in Fig. 2. 8 based on Task 1 and Task 2.*

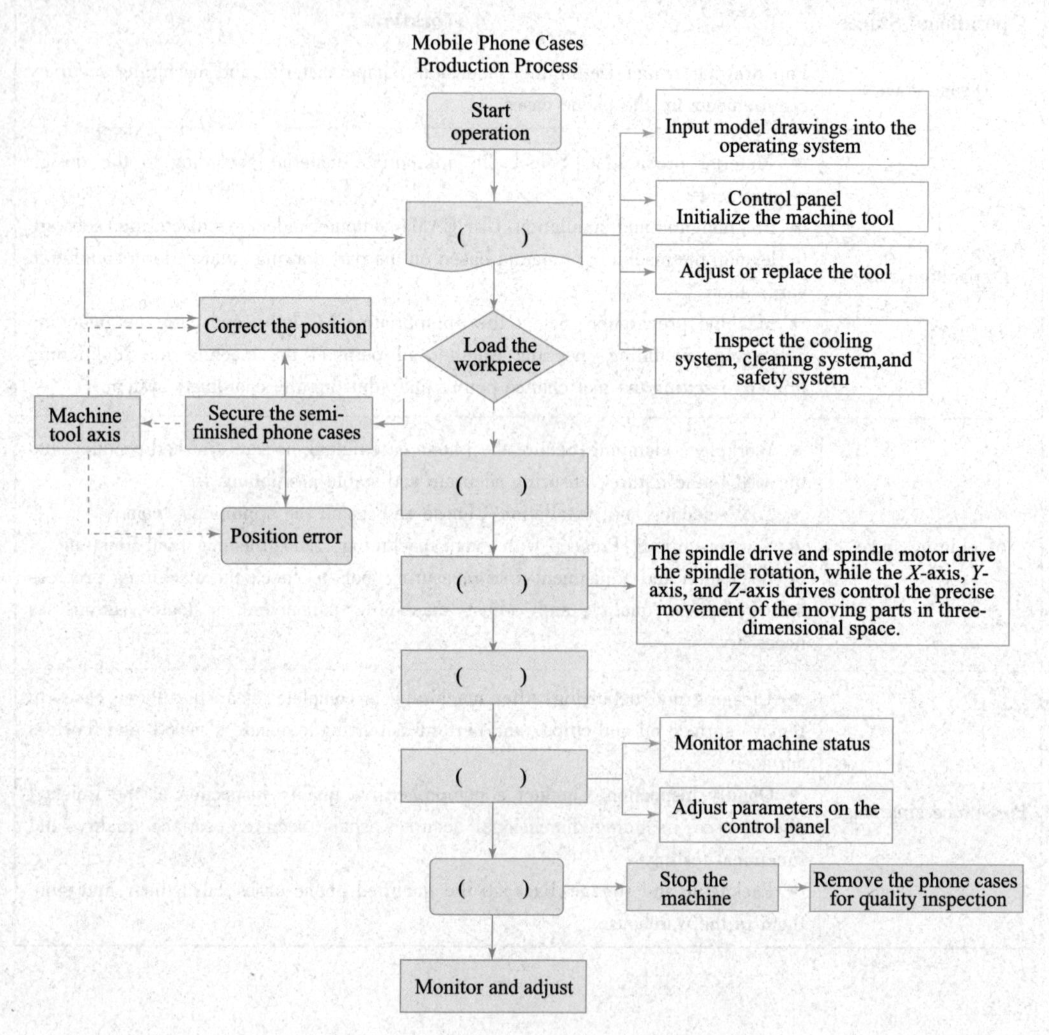

Fig. 2. 8 Mobile phone case production flowchart

Unit 3
Data in Intelligent Manufacturing
智能制造中的数据

Unit Introduction

Driving Questions

驱动问题

In intelligent manufacturing, data serves as the driving force behind innovation and efficiency, often referred to as the new "oil". Once connected, data can optimize every aspect of manufacturing. Big data platforms enable the collection, storage and analysis of vast amounts of production data. But what specific advantages can data bring to intelligent manufacturing? And how do big data platforms empower manufacturers to leverage these data-driven opportunities?

在智能制造中,数据是推动创新和提升效率的核心动力,常被称为新的"石油"。数据连接后,可以优化制造的方方面面。运用大数据平台能高效地收集、存储和分析海量的生产数据。那么,数据究竟为智能制造带来哪些好处呢?要把握数据驱动带来的发展机遇,该如何利用好大数据平台呢?

Project Overview

项目概述

In this unit, you will begin by exploring the basic concepts of industrial big data. After that, you will delve into essential tools such as data warehousing and data mining. Next, you will learn about the role of 5G technology in big data processing. Then, you will study practical application cases of how big data platforms are applied in China's manufacturing industry and the benefits they bring. Next, working in groups, you will design a simple brochure for ABC Company's big data platform. After presenting the brochure, you will discuss the practical applications and challenges of big data in intelligent manufacturing.

在本单元,首先学习工业大数据的基本概念,然后深入学习数据仓储和数据挖掘等重要工具,了解5G技术在大数据处理中的作用。之后,研究中国大数据平台在制造业中的实际应用案例及其带来的好处。接下来,小组合作为ABC公司的大数据平台制作一份宣传册,并在班级展示宣传册。最后,讨论大数据在智能制造中的实际应用和挑战。

Project Extension

项目拓展

After completing the main project, you'll engage in a hands-on extension activity to deepen your understanding of an industrial cloud platform. This will involve learning how to monitor and manage such a platform.

完成主要项目后,将进行实操拓展,学习如何监控和管理工业云平台,以进一步巩固对工业云平台的理解。

Learning Objectives

学习目标

This unit is intended to help you:

1. have a general idea of industrial big data;
2. explore the key tools and methods of big data processing;
3. analyze cases of big data platforms in simple English;
4. design and present an English brochure for big data platforms;
5. monitor and manage an industrial cloud platform;
6. cultivate awareness of data ethics.

本单元旨在帮助你:

1. 了解工业大数据;
2. 探讨大数据处理的主要工具和方法;
3. 用英语简单分析大数据平台的应用案例;
4. 制作并展示大数据平台的英文宣传册;
5. 监控和管理工业云平台;
6. 培养数据伦理意识。

Getting Ready

▶ What is a brochure, and what are the linguistic features of its text? Please complete the tasks below to find the answer. Fig. 3.1 is a photo of a brochure.

什么是宣传册?宣传册文本有什么特点?请完成以下任务,找到答案。图3.1是一张宣传册照片。

Fig. 3. 1 Photo of a brochure

Task 1 *Listen to the passage and fill in the blanks with the correct words.*

A brochure is a promotional document primarily used to introduce a 1. ________, organization, products, or 2. ________ and inform prospective customers or members of the public of the 3. ________. A brochure is a corporate 4. ________ instrument to promote a product or service. It is a 5. ________ used to circulate information about the product or service. A brochure is like a 6. ________ but with pictures of the product or the service which the brand is promoting. A brochure is usually 7. ________ and only includes promotional 8. ________ information.

Task 2 *Learn some linguistic features of brochures.*

产品宣传册中常常会用到从句,从句可以提供更多细节,帮助解释产品的特点和优势,使宣传语更具吸引力和说服力。

2.1 定语从句(Attributive Clause)

示例: This is a product **that** will change the way you work.

Our new device, **which** is designed with the latest technology, offers unparalleled performance.

2.2 状语从句(Adverbial Clause)

示例: We choose TechPro X200 **because** it delivers unmatched quality and performance.

Since it's crafted with premium materials, you can rely on its durability.

Task 3 *Fill in the blanks with words like that, which, because, since, as, who, when and so on.*

1. The software ________ we developed is tailored to optimize production processes.
2. We are expanding our facility ________ the demand for our products has increased.
3. Our latest model, ________ uses solar energy, is environmentally friendly.
4. ________ the new system is installed, our productivity will double.
5. We launched a new robot ________ can operate autonomously.
6. The CEO, ________ was appointed last year, has initiated several cost-cutting measures.
7. Our factory employs over 1000 workers ________ are highly skilled.
8. We will introduce a new assembly line ________ features the latest technology.
9. Our conveyor belts, ________ handle materials with exceptional care, are

crucial to our operations.

10. You will receive the order next week ________________ we will ship it tomorrow.

11. Our production capacity has doubled ________________ recent upgrades.

12. ________________ we implement these changes, our efficiency will improve significantly.

Starting up

▶ Data has always been a part of the manufacturing industry. However, it was only with the rise of big data that the manufacturing industry truly transformed. What is big data? And what is industrial big data? In this section, we'll learn some fundamental concepts of industrial big data.

数据一直是制造业的一部分,但是直到大数据的兴起,制造业才真正迎来了变革。那么,什么是大数据? 什么是工业大数据? 在这一部分,我们将学习工业大数据的一些基本概念。

Industrial Big Data

Definition

Industrial big data refers to the large amounts of different types of data produced rapidly by industrial machines, which are connected through IoT. The term became popular in 2012 alongside "Industry 4.0". Similar to "big data" in IT, industrial big data holds significant business potential because the data from these machines can be very valuable. By using Industrial Internet technology, this data helps in making better management decisions, cutting down maintenance costs, and improving customer service.

Big data refers to data generated with high volume, high variety, and high velocity, requiring new processing technologies for better decision making, knowledge discovery, and process optimization. Sometimes, veracity is also emphasized to highlight the quality and integrity of the data. However, industrial big data has additional characteristics. Beyond the usual "V's" of big data, it includes visibility and value. These extra "V's" make industrial big data distinct by emphasizing its unique capacity to reveal and create substantial business value from previously invisible insights.

Sources

From the perspective of data sources, industrial big data mainly comprises three categories:

➢ Business data related to enterprise operations and management

This category includes data from enterprise informatization domains such as enterprise resource planning (ERP), product lifecycle management (PLM), supply chain management (SCM), customer relationship management (CRM), and energy management systems (EMS). This type of data represents the traditional data assets of industrial enterprises.

➢ Manufacturing process data

This refers to the data generated during industrial production, including operational status parameters of equipment, materials, and product processing, as well as environmental parameters. These production condition data are transmitted in real-time through manufacturing execution system (MES). Given the extensive use of intelligent equipment, the volume of this data category is growing the fastest.

➢ External data

This includes data on the usage and operational status of industrial enterprise products after they have been sold. It also encompasses extensive lists of customers and suppliers, as well as external Internet data.

The Application Scenarios

As information technology and industrialization continue to deeply integrate, the data owned by industrial enterprises has become increasingly abundant. This includes design data, sensor data, automated control system data, production data, and supply chain data. The value driven by data and the insights it brings span the entire lifecycle of intelligent manufacturing. In recent years, new models and applications driven by the development of intelligent manufacturing and the Industrial Internet have further enriched the application scenarios of industrial big data. Typical application scenarios include intelligent design, intelligent production, networked collaborative manufacturing, intelligent services, and personalized customization.

New Words and Phrases

rapidly[ˈræpɪdli] *adv.* 快速地

significant[sɪgˈnɪfɪkənt] *adj.* 重大的

potential[pəˈtenʃ(ə)l] *n.* 潜力

valuable[ˈvæljuəb(ə)l] *adj.* 有价值的

maintenance[ˈmeɪntənəns] *n.* 维护

volume[ˈvɒljuːm] *n.* 体量,量

variety[vəˈraɪəti] *n.* 多样性

velocity[vəˈlɒsəti] *n.* 速度

optimization[ˌɒptɪmaɪˈzeɪʃn] *n.* 最优化

veracity[vəˈræsɪti] *n.* 真实性

emphasized[ˈɛmfəˌsaɪzd] *adj.* 强调的

highlight[ˈhaɪlaɪt] *v.* 突出,强调

integrity[ɪnˈtɛgrəti] *n.* 完整性

additional[əˈdɪʃənəl] *adj.* 额外的

characteristic[ˌkærɪktəˈrɪstɪk] *n.* 特征

visibility[vɪzəˈbɪlɪti] *n.* 可见性

capacity[kəˈpæsəti] *n.* 能力

reveal[rɪˈviːl] *v.* 揭示

substantial[səbˈstænʃ(ə)l] *adj.* 实质的

insight[ˈɪnsaɪt] *n.* 洞察力,深刻的理解

comprise[kəmˈpraɪz] *v.* 包含,由……组成

enterprise[ˈentərpraɪz] *n.* 企业

domain[dəˈmeɪn] *n.* 领域,范围

represent[ˌrɪprɪˈzɛnt] *v.* 代表,象征

asset[ˈæsɛt] *n.* 资产

abundant[əˈbʌndənt] *adj.* 丰富的

span[spæn] *v.* 跨越,包括

enrich[ɪnˈrɪtʃ] *v.* 丰富

customization[ˈkʌstəmaɪzeɪʃən] *n.* 定制,个性化

refer to　指的是

be similar to　与……相似

cut down　削减

be generated with　生成,产生

from the perspective of　从……角度来看

be related to　与……有关

Technical Terms

industrial big data　工业大数据

product lifecycle management(PLM) 产品生命周期管理

supply chain management(SCM) 供应链管理

customer relationship management(CRM) 客户关系管理

energy management systems(EMS) 能耗管理系统

Task 4 *According to the passage, are the following statements true (T) or false (F)?*

( )1. Industrial big data became popular alongside "Industry 4.0".

( )2. Industrial big data can help cut down maintenance costs and improve customer service.

( )3. Industrial big data does not emphasize the quality and integrity.

( )4. Industrial big data only has the traditional "V's" of big data, which are volume, variety, and velocity.

( )5. Business data related to enterprise operations and management is one of the main categories of industrial big data.

( ) 6. Manufacturing process data is transmitted in real-time through energy management systems.

( )7. Customer and supplier lists are considered part of external data.

( )8. Design data, sensor data, automated control system data, production data, and supply chain data are all owned by industrial enterprises.

Task 5 *Find the answers to the following questions in the passage.*

1. What is industrial big data?

____.

2. How does industrial big data differ from big data?

____.

3. What are the three categories of industrial big data sources?

____.

4. What does business data include in the text?

____.

5. List some common application scenarios of industrial big data.

____.

Task 6 *Match the English expressions in Column A with their Chinese meanings in Column B.*

Column A	Column B
1. industrial big data	A. 客户关系管理
2. supply chain management(SCM)	B. 产品生命周期管理
3. product lifecycle management(PLM)	C. 能耗管理系统
4. enterprise resource planning(ERP)	D. 企业运营管理
5. enterprise operations and management	E. 制造执行系统
6. customer relationship management(CRM)	F. 工业大数据
7. energy management systems(EMS)	G. 企业资源计划
8. manufacturing execution systems(MES)	H. 供应链管理

Task 7 *Make six sentences with words from the following boxes.*

volume value variety veracity

visibility velocity capacity integrity

Example: The **volume** of raw materials used in the manufacturing process has been optimized to reduce waste.

1. ______________________________.
2. ______________________________.
3. ______________________________.
4. ______________________________.
5. ______________________________.
6. ______________________________.

Task 8 *Translate the following sentences into English using the words and expressions given.*

1. 精益生产又称精益制造，重点是提高效率和减少浪费。(refer to)

____.

2. 政府鼓励居民少开汽车，以改善空气质量。(cut down)

____.

3. 预计新生产线将带来可观的收入。(generate with)

____.

4. 她和导师的处事风格相似，都注重团队协作。(be similar to)

____.

5. 阅读书籍可以丰富思想，拓宽视野。(enrich)

____.

Investigating

▶ Data warehousing and data mining are essential tools for modern data management and analysis. In this section, we will explore the characteristics and benefits of these tools.

数据仓储和数据挖掘是现代数据管理和分析的重要工具。在这一部分，我们将详细探讨这些工具的特点和优势。

Data Warehousing and Data Mining

Data warehousing is constructed to support management functions, whereas data mining is used to extract useful information and patterns from data. Data warehousing involves compiling information into a data warehouse.

Data Warehousing

Data warehousing is a technology that gathers structured data from multiple sources for comparison and analysis, rather than transaction processing. A data warehouse supports management decision-making by providing a platform for data cleaning, integration, and consolidation. A data warehouse contains subject-oriented, integrated, time-variant, and non-volatile data. It ensures data quality, consistency, and accuracy while consolidating data from various sources.

The advantages of data warehousing are as follows.

- Data warehousing make corporate data easier to understand.
- Data warehousing allow for frequent updates, keeping organizations current with their target audiences and customers.
- Data warehousing make data more accessible to businesses and organizations.
- Data warehousing store large volumes of historical data, enabling users to analyze trends and make future predictions.

Data Mining

Data mining is the process of finding patterns and correlations within large data sets to identify relationships between data (AI), statistics, database, and machine learning systems are all used in data mining technologies. Data mining employs several key techniques such as classification, regression, clustering, association rule learning, and anomaly detection to extract valuable insights from data.

The advantages of data mining are as follows.

- Data mining helps in analyzing and sorting data efficiently.
- Data mining quickly identifies and addresses system faults, eliminating potential dangers early.
- Data mining is more cost-effective and efficient than other statistical methods.
- Data mining helps companies provide valuable and easily accessible knowledge-based data.
- Data mining excels in detecting and identifying faults within systems.

Table 3. 1 is the comparison between data warehousing and data mining.

Table 3. 1 Comparison between data warehousing and data mining

No.	Aspects of Comparison	Data Warehousing	Data Mining
1	Process	Data is stored periodically.	Data is analyzed regularly.
2	Purpose	To extract and store data for easier reporting	To use pattern recognition algorithms to identify patterns
3	Feature	Subject orientation, data integration, time variance and data persistence	Use of AI, statistics, database, and machine learning systems

Continued

No.	Aspects of Comparison	Data Warehousing	Data Mining
4	Application	Organizes and stores data, making reporting easier and faster	Uses pattern recognition tools to identify access patterns
5	Use case	Connecting a data warehouse with operational systems like CRM to add value.	Creating patterns of key parameters, such as customer purchasing behavior and sales to help business adjust their operations and production

New Words and Phrases

construct[kən'strʌkt] *v.* 构建

extract[ɪk'strækt] *v.* 提取

pattern['pætən] *n.* 模式

compile[kəm'paɪl] *v.* 汇编

gather['gæðə(r)] *v.* 收集

multiple['mʌltɪp(ə)l] *adj.* 多重的

comparison[kəm'pærɪsən] *n.* 比较

transaction[træn'zækʃ(ə)n] *n.* 交易

consolidation[kənˌsɒlɪ'deɪʃ(ə)n] *n.* 合并

consistency[kən'sɪstənsi] *n.* 一致性

simplify['sɪmplɪfaɪ] *v.* 简化

corporate['kɔːpərət] *adj.* 公司的

update[ʌp'deɪt] *v.* 更新

frequent['friːkwənt] *adj.* 频繁的

current['kʌrənt] *adj.* 当前的

accessible[ək'sesəb(ə)l] *adj.* 易接近的

prediction[prɪ'dɪkʃ(ə)n] *n.* 预测

correlation[ˌkɒrə'leɪʃ(ə)n] *n.* 相关性

database['deɪtəˌbeɪs] *n.* 数据库

employ[ɪm'plɔɪ] *v.* 使用

technique[tek'niːk] *n.* 技术

classification[ˌklæsɪfɪ'keɪʃ(ə)n] *n.* 分类

regression[rɪ'grɛʃ(ə)n] *n.* 回归

clustering[ˈklʌstərɪŋ] n. 聚类

association[əˌsəʊsiˈeɪʃ(ə)n] n. 联合

anomaly[əˈnɒməli] n. 异常

sort[sɔːt] v. 排序

eliminate[ɪˈlɪmɪneɪt] v. 消除

excel[ɪkˈsel] v. 擅长

Technical Terms

data warehousing　数据仓储

data mining　数据挖掘

association rule learning　关联规则学习

anomaly detection　异常检测

Task 9 *Read the passage and match the two parts of the sentences.*

1. Data mining is	A. classification, regression, clustering, association rule learning, and anomaly detection to extract valuable insights from data.
2. Data warehousing is	B. used to extract useful information and patterns from data.
3. A data warehouse supports	C. users to analyze trends and make future predictions.
4. Data warehouses make	D. a technology that gathers structured data from multiple sources for comparison and analysis.
5. Data warehousing enables	E. system faults, eliminating potential dangers early.
6. Data mining employs	F. corporate data easier to understand.
7. Data mining helps	G. management decision-making by providing a platform for data cleaning, integration, and consolidation.
8. Data mining identifies and addresses	H. in analyzing and sorting data efficiently.

Task 10 *Translate the sentences that you completed in Task 9. The first one is provided.*

1. Data mining is used to extract useful information and patterns from data.
数据挖掘用于从数据中提取有用的信息和模式。

2.

3.

4.

5.

6.

7.

8.

Task 11 *Listen to a recording about big data and 5G. Fill in the blanks with the words provided in the box. And then read it aloud.*

• downtime	• minimizing	• optimize	• latency
• sensors	• actionable	• feedback	• cutting

5G and big data are key parts of IoT revolution. And for the simple reason that most IoT devices exist to produce 1. ________ data. The use of 5G and big data with AI-based machine vision is expected to turbocharge factories, helping to 2. ________ assembly lines, reduce 3. ________, and enable preventive maintenance. Using IoT 4. ________, AI-based predictive analytics can optimize inventory management, 5. ________ costs and 6. ________ delays. All this data processing should not only enhance and optimize business processes, but provide 7. ________ into the intelligent networks for continuous, self-healing optimization. The data-driven organization of the future features data processing close to its source, either through on-device AI or on-the-edge cloud, all connected via 5G. The performance needed for this intelligent organization comes from low 8. ________ enabled by 5G and the edge network.

Researching

► How are big data platforms being applied in China? Let's study and analyze some cases together.

中国的大数据平台是如何应用的？让我们一起研究和分析一些案例。

全球领先的IT市场研究和咨询公司IDC发布的《中国大数据平台市场份额,2023:数智融合时代的真正到来》(2024年8月)报告显示,2023年中国大数据市场相比2022年增长24.6%,尤其在先进制造、汽车、金融等行业具备强劲增长潜力,整体市场空间仍在高速增长。

Task 12 *Cases Study: The Main Application of Big Data Platforms in Manufacturing in China (Fig. 3.2).*

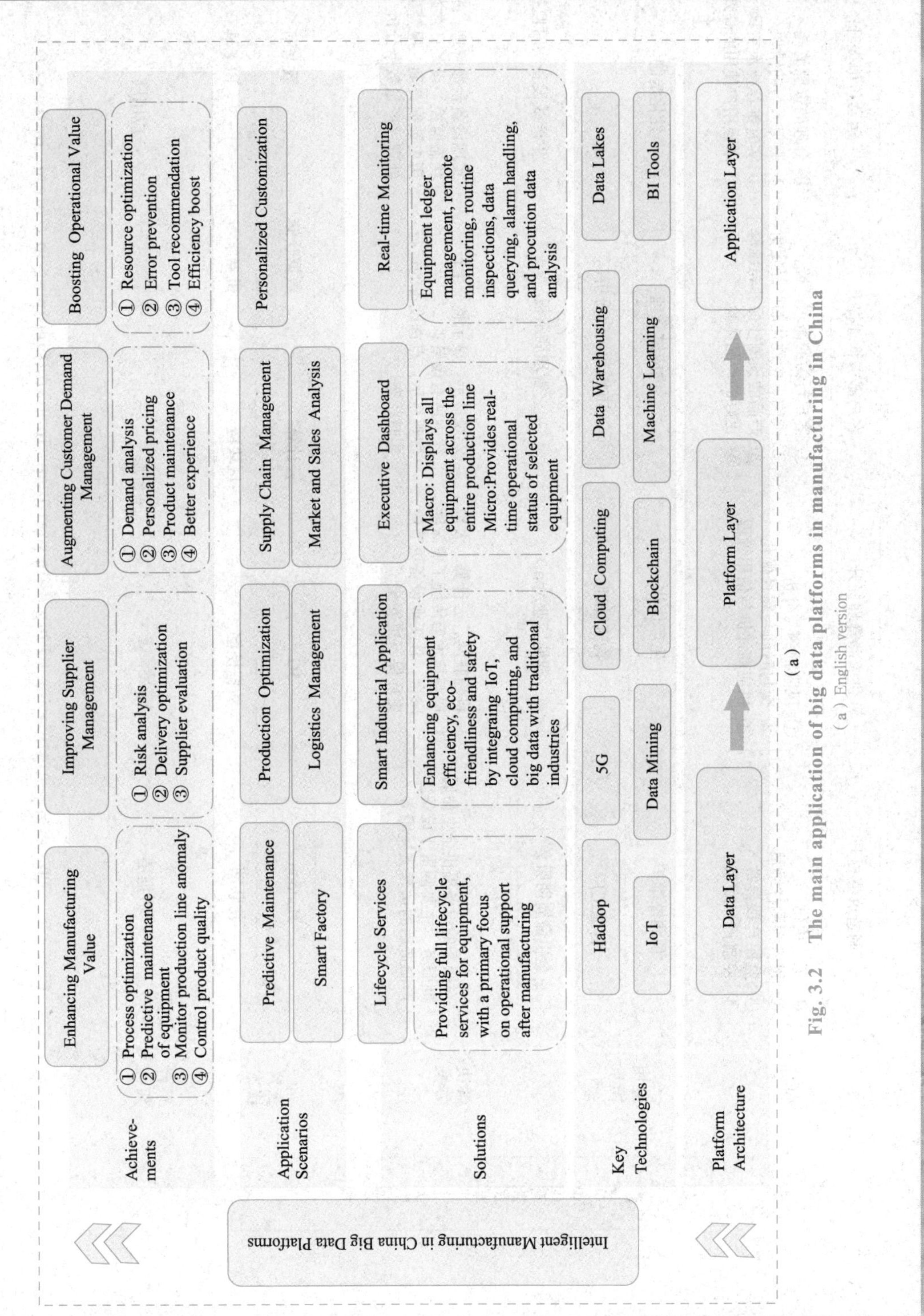

(a)

(a) English version

Fig. 3.2 The main application of big data platforms in manufacturing in China

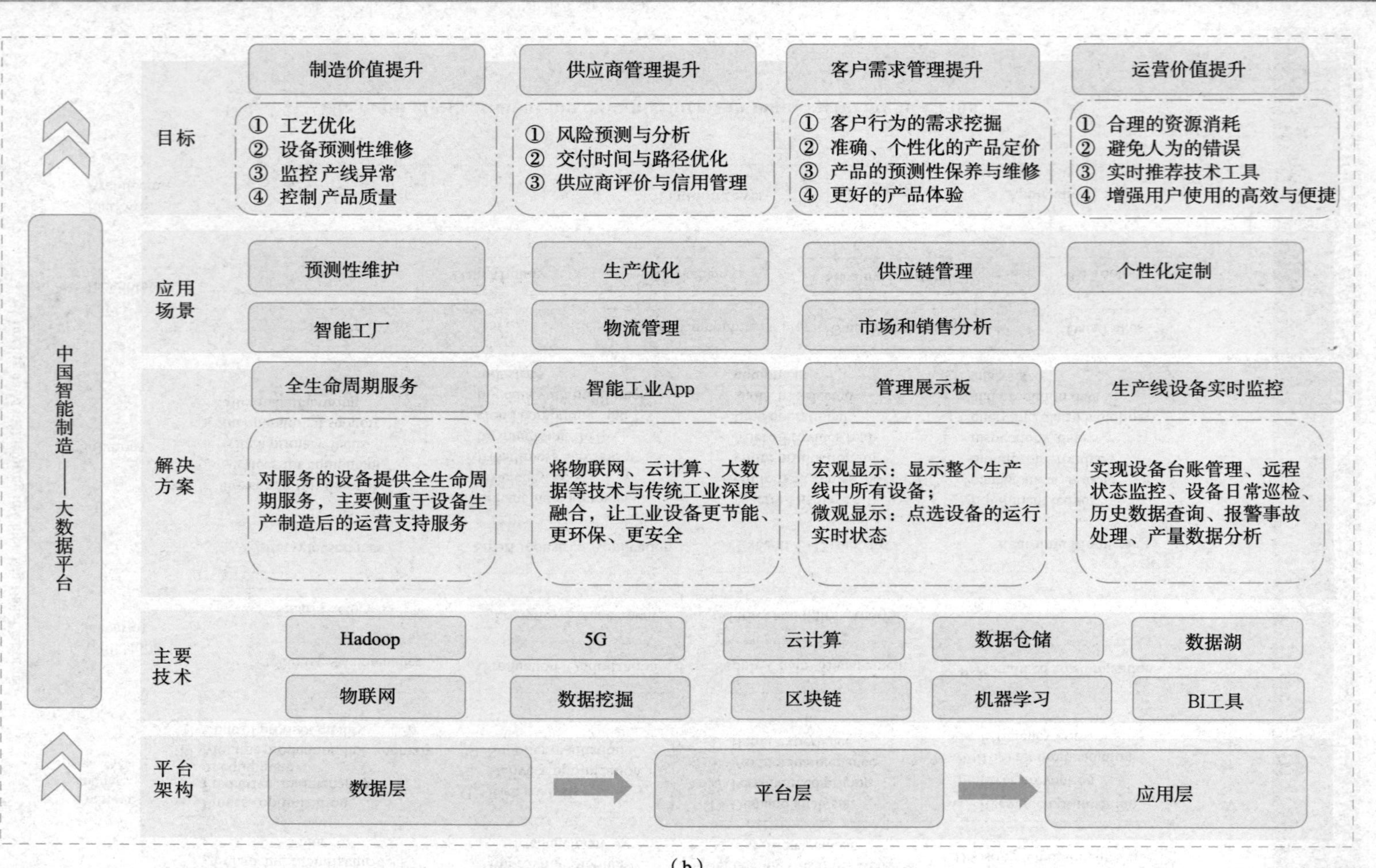

(b)

Fig. 3.2 The main application of big data platforms in manufacturing in China(Continued)

(b) Chinese version

Task 13 ***Cases Analysis: Analyze the key technologies, scenarios, solutions and achievements of the big data platforms of a particular Chinese company.***

Building and Presenting

Building

实施

▶ You need to design a simple brochure for ABC Company's big data platforms. As a group of 4 – 6 members, you may utilize AI technology, online research, fieldwork, and the knowledge gained from this unit to better understand the topic. Then work through the following steps in Fig. 3. 3 to help you complete the project.

制作一份 ABC 公司大数据平台的宣传册。请以小组(4 ~ 6 名成员)为单位,通过使用 AI 技术、网络调研、实地考察等方法,结合本单元所学知识,深入了解主题,并按照图 3. 3 中的步骤流程完成宣传册制作。

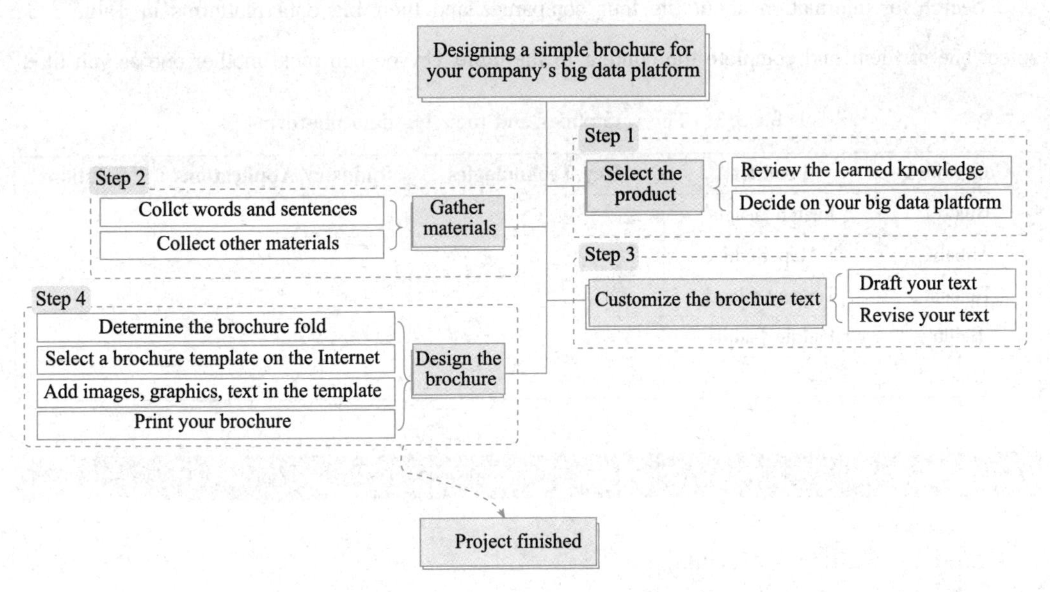

Fig. 3. 3　Brochure creation flowchart

Task 14

Step 1 Select the Product

A. Review the learned knowledge

Answer the following questions in Table 3. 2 based on what you've learned before.

Table 3. 2 List of questions under three different topics

Topics	Questions	Notes
Industrial big data	What is industrial big data?	
	What are the major benefits?	
Data warehousing and data mining	What is data warehousing?	
	What is data mining?	
	What are the functions?	
Big data platform	What is big data platform?	
	What technologies are used?	
	What are the advantages?	

B. Decide on your big data platform

Search for information about the four companies and their big data platforms in Table 3. 3, select one of them, and complete the content in the table. Or you can pick another one as you like.

Table 3. 3 The companies and their big data platforms

Companies	Product	Key Technologies	Industry Applications	Solutions
Huawei	Fusion Insight			
Alibaba	E-Map Reduce			
Tencent	Tencent Cloud			
Baidu	Paddle Paddle			
Others				

Task 15

Step 2 Gather Materials

A. Collect words and sentences

Brainstorm words or expressions related to big data platform (Fig. 3. 4) , and then make sentences with them.

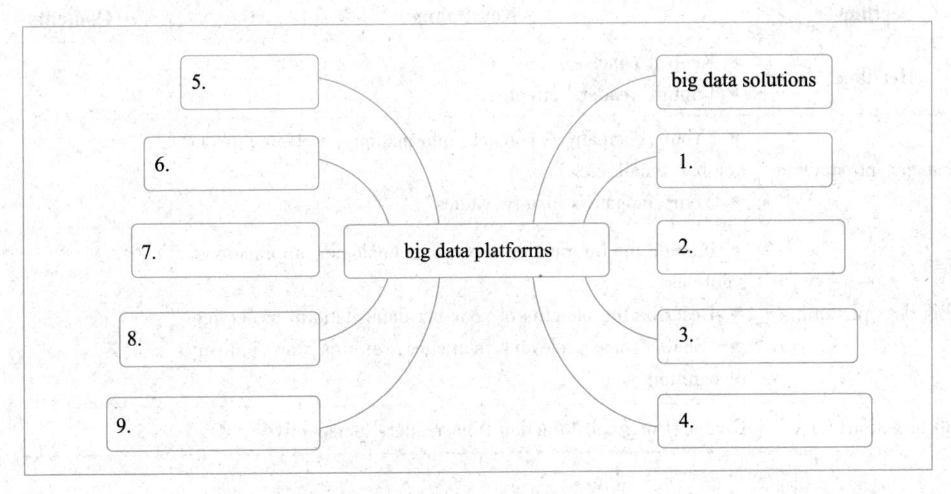

Fig. 3.4 Words and expressions related to big data platforms

Example: With Fusion Insight, Huawei has built a hugely powerful collection of big data solutions that can create greater value for customers and significantly enhance the efficiency of work and life.

1.

2.

3.

4.

5.

6.

B. Collect other materials

Search for and collect images and graphics you're planning to include.

Task 16

Step 3 Customize the Brochure Text

A. Draft your text

Draft the text for your brochure of the big data platform according to Table 3.4, making sure it covers the key points in Task 14.

Table 3.4 Sections of a brochure and writing requirements

Sections	Key Points	Contents
Headline	• Keep it concise • Capture readers' attention	
Company introduction	• Your company's contact information (website, phone number, email, etc.) • Your company's history, values	
Big data platform	• Include the information such as technologies, applications, solutions • Describe the benefits of your big data platform or service • Show some relevant statistics, awards, or industry recognition	
Call to action(CTA)	Give a strong call to action from readers' perspective	

B. Revise your text

Polish your brochure text according to the language skills you learned in Task 2 (Table 3.5).

Table 3.5 Comparison of wording in the text and Wording after polishing

Wording in the text	Wording after polishing

Task 17

Step 4 Design the Brochure

After finishing your text of brochure, follow the steps in Table 3.6, and mark each one as accomplished as you go, until you finish your brochure.

Table 3.6 Brochure-making Process

Brochure-making Process	
• Customize your brochure text	(√)
• Determine the brochure fold	( )
• Select a brochure template on the Internet	( )

Continued

Brochure-making Process	
• Add images, graphics, text in the template	( )
• Print your brochure	( )

Presenting

展示

▶ Now that you have made your brochure, it is time to share your work with your classmates.

宣传册制作完成后,请与同学分享你们的成果。

Task 18 *As a group, present your brochure in class. According to the schedule in Table 3.7, take questions, comments, or suggestions from the audience if possible.*

Table 3.7 Task assignment

Leader	Student A	Student B	Student C	Student D
Opening and ending	Theme	Linguistic skills	Technical strategies	Artistic strategies

Task 19 *The group leader sends the modified brochures to the teacher's mailbox or uploads them to the online platform as the usual performance evaluation document and for everyone to learn from online.*

Assessing and Reflecting

Assessing

评估

▶ Assessing the work you have accomplished allows you to know not only the response of your audience to your work but also the weak parts of your work to improve.

评估你所完成的工作，不仅能了解观众对你作品的反馈，还能发现不足，以便改进。

Task 20 ***The group project is evaluated jointly by teachers and students (Table 3. 8).***

Table 3. 8 Group project evaluation form

Evaluation (*Excellent* = 3, *Good* = 2, *Poor* = 1)					
Creativity	**Language**	**Creativity**	**Presentation**	**Collaboration**	**Cross-subject (brochure)**
Group1					
Group2					
Group3					
Group4					
Group5					

Reflecting

反思

► Reflecting on your learning helps you understand what you've gained from this unit and how you can apply it in the future.

反思有助于你理解在本单元中所学的知识，以及将来如何加以运用。

Task 21 ***The following questions in Table 3. 9 may help you do the reflection, but feel free to ask yourself more questions when necessary.***

Table 3. 9 Questions reflection form

Questions	Reflections
Did I increase my vocabulary to finish my task more precisely?	
Did I use any digital technologies?	
Did I actively engage in preparing the project?	
Did I develop my presentation skills through presenting my group's ideas and work?	
Did my contributions help others understand data in IM better?	
Did I identify, analyze and solve problems by fulfilling various tasks?	

Extending

▶ You are a technician at a large glass manufacturing group. In your daily work, you identified an abnormal energy consumption warning via the industrial cloud platform. Now, you need to fill in a work order on this issue and send the warning to the production line through the platform so that the production team can address the problem quickly.

你是某大型玻璃生产集团的技术人员，在日常工作中，你通过工业云平台发现了能耗异常预警。你需要根据异常情况填写一份事件工单，并通过此平台将该异常信息推送到生产线，以便快速解决问题。

Task 1 *Get acquainted with the industrial cloud platform.*

The industrial data cloud platform enables the analysis of industrial data, the construction of a data asset system, and the establishment of workshop evaluation metrics such as operating conditions, energy consumption, and production output, ultimately optimizing the production process. The industrial cloud platform consists of two main components: the industrial equipment cloud and the intelligent transformation of traditional equipment, as shown in Fig. 3. 5.

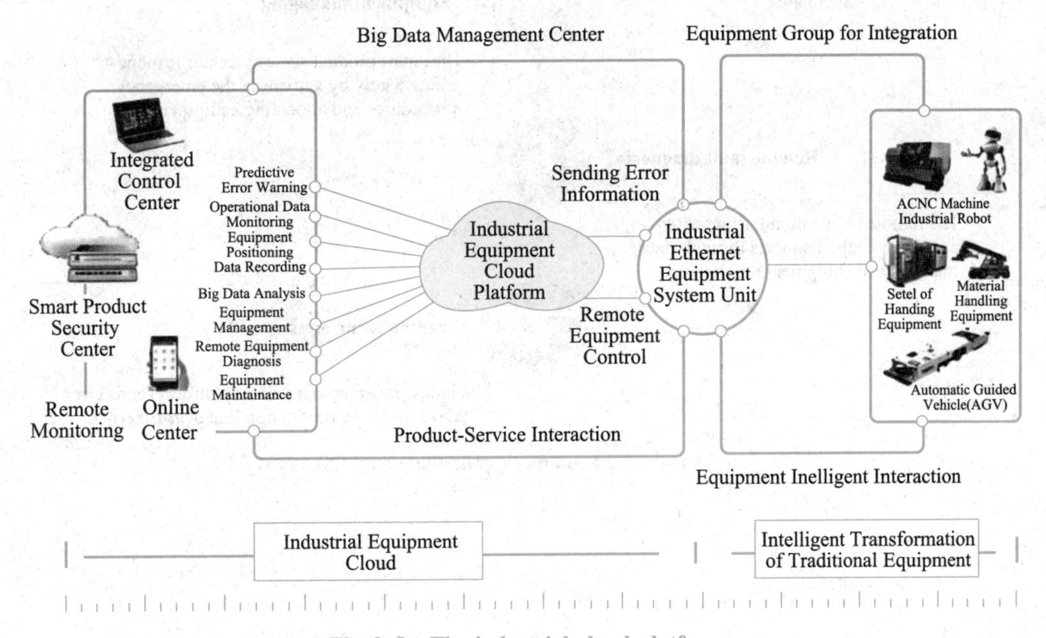

Fig. 3. 5 The industrial cloud platform

Task 2 *Familiarize yourself with the workflow for managing production anomalies via the industrial data cloud platform (Fig. 3. 6).*

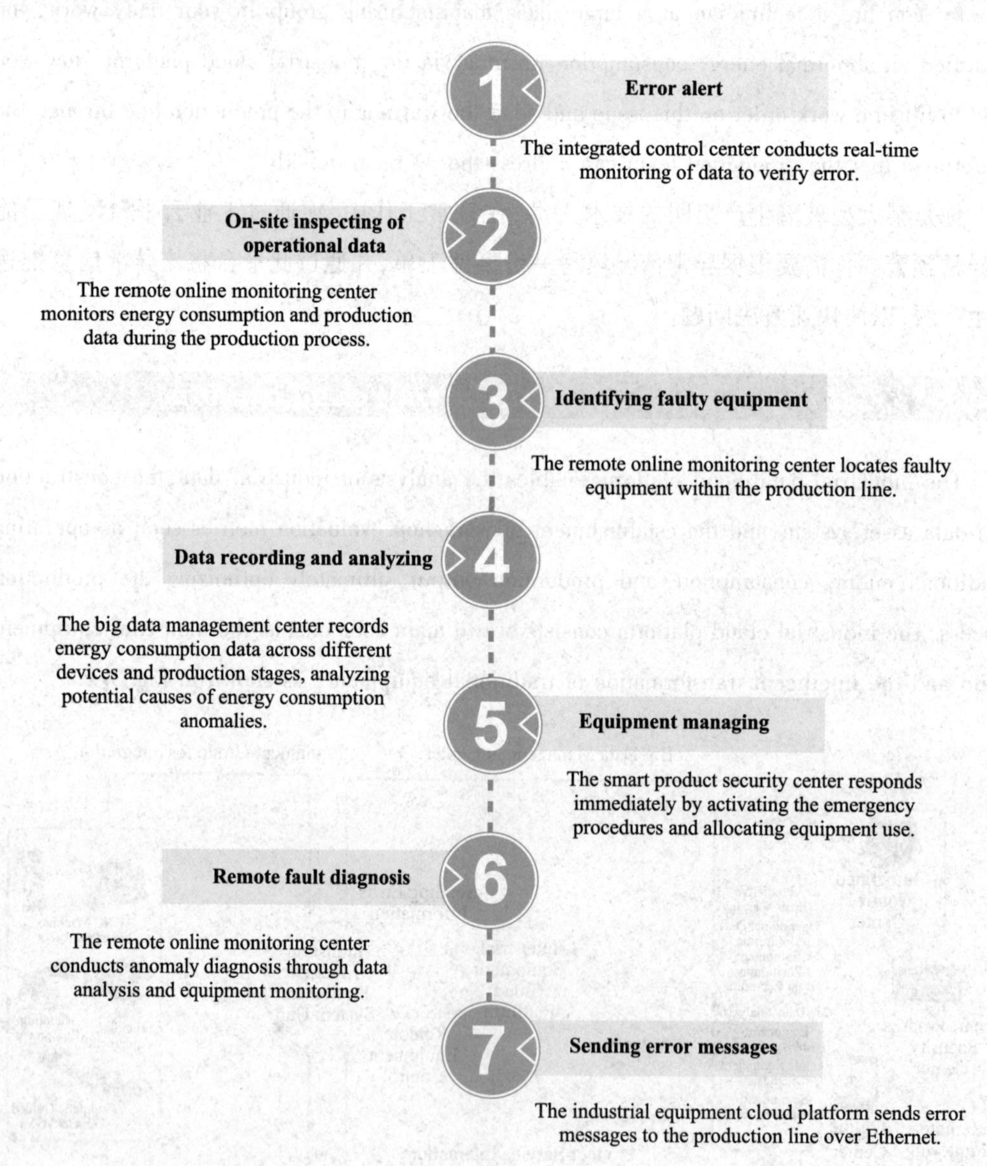

Fig. 3. 6 Production anomalies management flowchart

Task 3 *Finish the work order (Table 3.10) based on Task 1 and Task 2.*

Table 3.10 Integrated control center work order

<table>
<tr><th colspan="4">Integrated Control Center Work Order</th></tr>
<tr><td>Recorder</td><td>Li Lei</td><td>Date of Record</td><td>July 10, 2022</td></tr>
<tr><td>Subject</td><td colspan="3">Analysis and handling of energy consumption anomaly on the glass production line</td></tr>
<tr><td>Recipient</td><td>Team Leader Zhao</td><td>Department</td><td>Production Department</td></tr>
<tr><td>1. ____________</td><td colspan="3">The glass production line CX002 triggered an energy consumption error warning. The system automatically activated the anomaly alert system, indicating that there is an unusual energy consumption on the current production line, requiring immediate verification and resolution.</td></tr>
<tr><td>2. ____________</td><td colspan="3">• Energy consumption data:
Current total energy consumption: 120 kW · h/t, exceeding the standard limit by 7 kW · h/t.
Equipment in high energy consumption: the furnace consumes 60 kW · h/t of energy, accounting for 50% of total energy consumption; the cutter consumes 30.6 kW · h/t of energy, taking up 25.5%.
• Production parameters:
Furnace temperature: 1,100 ℃, deviating 100 ℃ from the target temperature.
Production line energy consumption: 860 t/d, higher than the expected energy consumption (500 – 800 t/d).</td></tr>
<tr><td>3. ____________</td><td colspan="3">The monitoring system quickly located the issue in the furnace area, specifically in the heating element B zone, where significant temperature fluctuations were observed.</td></tr>
<tr><td>4. ____________</td><td colspan="3">Detailed data within one hour before and after the anomaly have been automatically recorded, including but not limited to energy consumption, temperature, pressure, and flow rates of various equipment. An anomaly report has been generated (Report No. BG112).</td></tr>
<tr><td>5. ____________</td><td colspan="3">Emergency procedures were immediately activated, adjusting the production line to low-load operation to reduce energy losses.</td></tr>
<tr><td>6. ____________</td><td colspan="3">Through the industrial equipment cloud platform, a remote connection was established to the furnace control system for an in-depth analysis of the control logic and sensor data in heating element B zone.</td></tr>
<tr><td>7. ____________</td><td colspan="3">This work log has been sent to the production engineers and relevant personnel instantly via email and App notifications, ensuring information spreads quickly and responses are fast.</td></tr>
</table>

Unit 4 Integration in Intelligent Manufacturing

智能制造中的集成

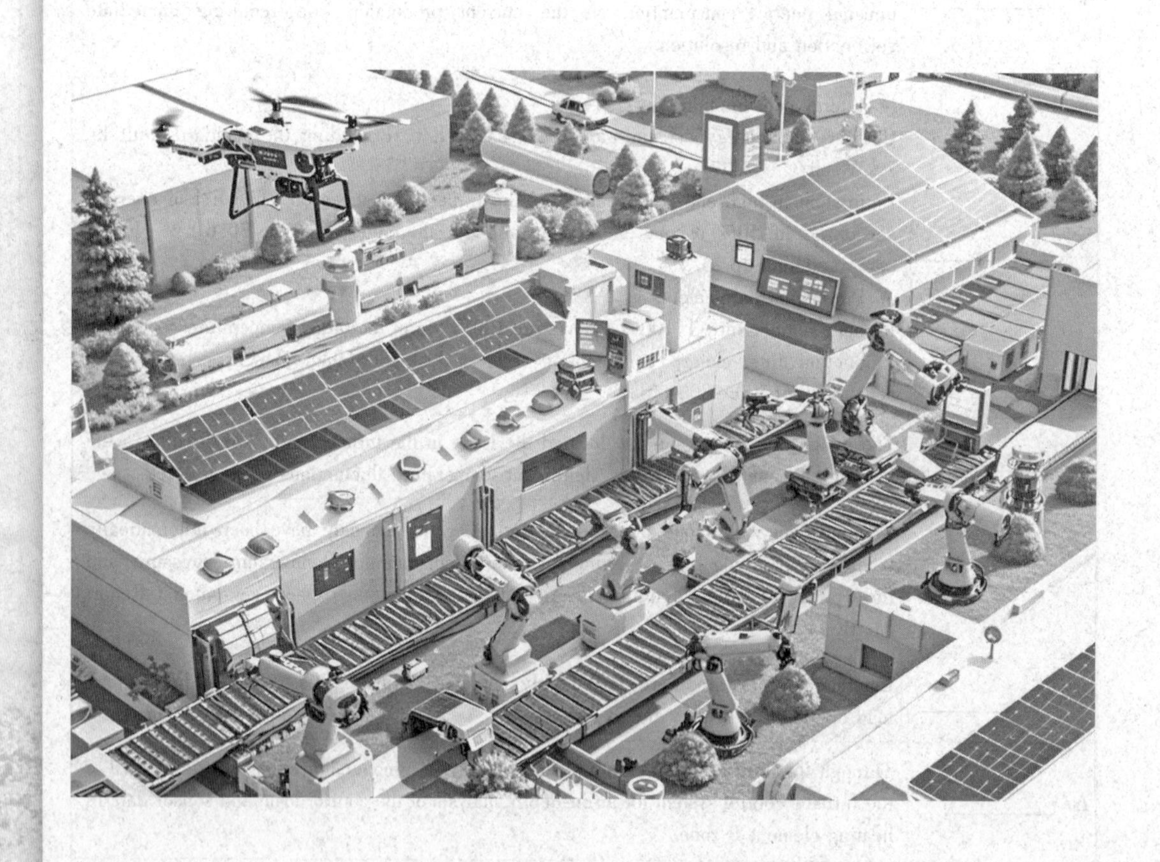

Unit Introduction

Driving Question

驱动问题

In intelligent manufacturing, integration refers to the seamless combination of various stages of the production process, different hierarchical levels, and the entire supply chain to achieve high levels of automation and optimization. As pioneers of intelligent manufacturing, lighthouse factories represent the highest level of integration. So, how do the lighthouses achieve efficient production and quality improvement through integration strategies?

在智能制造中,集成是指将生产过程的各个环节、不同层级以及整个供应链有机结合,以实现高度自动化和优化的生产过程。灯塔工厂作为智能制造的先锋,代表了最高水平的集成。那么,灯塔工厂是如何通过集成策略实现高效生产和质量提升的呢?

Project Overview

项目概述

In this unit, you will start by learning the basic concepts and challenges of horizontal integraton and vertical integration in manufacturing. Then, you will explore the concept of Industrial Internet Platforms(IIP) and how they help solve integration challenges, with a focus on the RootCloud platform as an example. After that, you will learn about the basics of MES and their role in supporting manufacturing integration. Next, you will study how China's lighthouse factories lead in manufacturing innovation through advanced integration strategies. Working in groups, you will participate in a role-play on a tour of guiding customers to ABC lighthouse factory. After the role-play, you will discuss how integration strategies can be applied to further enhance intelligent manufacturing processes.

在本单元,首先学习智能制造中横向集成和纵向集成的基本概念及其面临的挑战。然后,深入学习工业互联网平台的基本概念,重点以树根互联的根云平台为例说明工业互联网平台如何解决集成问题。接下来,了解 MES 系统的基础知识及其在支持制造集成中的作用。之后,研究中国的灯塔工厂如何通过先进的集成策略引领制造创新。随后,小组合作设计一个陪同客户参观 ABC 灯塔工厂的角色扮演。最后，讨论集成策略如何提升智能制造的全过程。

Project Extension

项目拓展

After completing the main project, you'll engage in a hands-on extension activity to deepen your understanding of lighthouse factories. This will involve learning how to draft a research plan about such a factory.

完成主要项目后，将进行实操拓展并填写灯塔工厂调研报告，以进一步巩固对灯塔工厂的理解。

Learning Objectives

学习目标

This unit is intended to help you:

1. have a general idea of integration in manufacturing;
2. explore the concept of IIP and their role in addressing integration challenges;
3. analyze cases of Chinese lighthouse factories in simple English;
4. participate in a role-play of a lighthouse factory tour in English;
5. draft a research plan;
6. strengthen awareness of collaboration within and outside the team.

本单元旨在帮助你：

1. 了解制造中的集成；
2. 探讨工业互联网平台的定义及其在应对集成挑战中的作用；
3. 用英语简单分析中国灯塔工厂的案例；
4. 完成灯塔工厂参观的英文角色扮演；
5. 填写一份调研报告；
6. 增强团队内外的合作意识。

Getting Ready

▶ How to accompany customers on a factory tour (Fig. 4. 1)? What are the characteristics of the language used when communicating with them? Please complete the tasks below to find the answer.

怎样陪同客户参观工厂(图4.1)? 与客户沟通时的用语有什么特点? 请完成以下任务,找到答案。

Fig. 4. 1 Accomany customers on a factory tour

Task 1 *Listen to a recording about a factory tour and answer the following questions.*

1. What kind of people typically participate in a plant tour?

Participants in a plant tour can include ________, ________, ________, ________.

2. What is the purpose of a plant tour?

The purpose of a plant tour is to provide a group of people, including customers, investors, stakeholders, or top management, with a guided walking tour of the manufacturing plant. This allows them to ________, ________, ________ knowledge about the manufacturing processes.

3. Why are plant tours beneficial for a company?

Plant tours are beneficial for a company as they help ________, ________, ________ and create greater confidence in your capacity to deliver both product quality and quantity.

4. How does a plant tour differ from traditional methods of showcasing a company's capabilities?

A plant tour differs from traditional methods like spreadsheets and presentations by offering a direct, immersive experience. It allows participants to obtain ________ and ________ understanding of how products are made and how the production line works.

5. What aspects of the company can a plant tour highlight?

A plant tour can highlight various aspects of the company, including the ________, ________, and amazing staff.

Task 2 *Learn some linguistic features on receiving customers.*

在接待客户时的交流对话中,语言的使用应平衡寒暄交谈和信息传达。

2.1 开放性问题(Open Questions)

特点:鼓励客户的互动和参与。

示例: What...? Why...? Can you tell me about...?

2.2 情态动词(Modal Verb)

特点:表示礼貌。

示例: Could we stop here for a moment?

Would you like to see how this machine works?

2.3 祈使句(Imperative Sentence)

特点:给出明确的、直接的指令,确保安全和正确指引。

示例: Please follow me.

Stay behind the yellow line.

2.4 现在进行时(The Present Continuous Tense)

特点:有效描述正在发生的动作或事件。

示例: We are currently moving towards the next section.

2.5 一般现在时(The Present Tense)

特点:用于描述事实或操作标准。

示例: We focus on sustainability.

Task 3 *Please rewrite the following sentences according to the requirements.*

Convert the following yes/no questions into open questions.

1. Do you like this product?

____?

2. Are you happy with our services?

____?

3. Is this your first time here?

____?

Rewrite these instructions using imperatives.

1. I think you should follow me.

____

2. It would be a good idea to stay behind the yellow line.

____

3. You should look at the monitor.

____

Make the following request /suggestions more polite using modal verbs.

1. Wait here.

____

2. Show them the machine.

____

3. Take a seat.

____

Fill in the blanks with the appropriate tense.

1. Right now, we ____(move) towards the quality control section.

2. Our factory ____(produce) 1,000 units every day.

3. It usually ________(take) them a few days to process and ship the order.

4. They ________(hold) a product development meeting in the main conference room right now.

Starting up

▶ Two of the most important strategies for production optimization are horizontal integration and vertical integration. What do you already know about integration? In this section, we will define them and explore their common challenges.

在智能工厂,优化生产的两大重要策略是横向集成和纵向集成。对于集成你了解多少? 在本部分,我们将会探讨这两种集成的定义及其所面临的挑战。

Horizontal Integration and Vertical Integration in Intelligent Manufacturing

Imagine a company as a tree, its branches reaching horizontally and its roots growing deeper vertically. Horizontal integration expands the company's reach, while vertical integration strengthens its foundation.

Horizontal Integration and Vertical Integration in Manufacturing

In manufacturing, horizontal integration refers to networking between individual machines, equipment, or production units. Vertical integration connects beyond traditional production hierarchy levels, from sensors to the business level.

Horizontal integration ensures that machinery, IoT devices, and engineering processes work together seamlessly. Vertical integration allows production data to be used for business, staffing, and other decisions by facilitating communication between the horizontally integrated shop floor and systems like ERP.

By combining horizontal integration and vertical integration, businesses can achieve end-to-end integration, where information and processes flow seamlessly across all levels and functions. This comprehensive integration enables real-time, data-driven decision-making, optimizes resource utilization, and enhances operational efficiency. Ultimately, it strengthens the company's ability to respond quickly and effectively to market changes and customer needs.

Horizontal Integration and Vertical Integration Challenges Manufacturers Face

Although horizontal integration and vertical integration can streamline production and enable smarter business decisions, the transition can pose some challenges.

- Lack of communication: Horizontal integration assumes a factory has devices and software capable of collecting and sharing production data. To achieve this integration, factories may need to purchase new IoT-enabled equipment or modify existing equipment to allow inter-connectivity.

- Knowledge silos: Breaking down data and knowledge silos is a challenging task. It begins on the production floor, where equipment and production units from diverse vendors offer varying levels of automation, are equipped with a wide range of sensors, and use different communication protocols. In other words, they often do not "speak the same language", requiring the establishment of a meta-network to bridge these communication gaps.

- Asset management: Once machines, data systems, and devices are integrated into your intelligent factory, ongoing management of these assets is essential to ensure security and productivity.

- Supply chain management: Whether connecting with external suppliers or integrating the internal supply chain, manufacturers must leverage data-driven technology to ensure smooth collaboration and effective decision-making.

- Data security: Horizontal integration requires sharing data with suppliers, subcontractors, partners, and often customers. This transparency enhances production agility and flexibility, but it also presents the challenge of ensuring all data is kept secure and accessible only on a need-to-know basis.

To solve these problems is never easy. However, the efforts usually pay off, as the integration results in more cost-effective production, consolidated management, and reduced overhead costs.

New Words and Phrases

branch[brɑːntʃ] *n.* 树枝

root[ruːt] *n.* 树根

horizontal[ˌhɒrɪˈzɒnt(ə)l] *adj.* 横向的;水平的

vertical[ˈvɜːtɪk(ə)l] *adj.* 纵向的;垂直的

integration[ˌɪntɪˈgreɪʃ(ə)n] *n.* 整合;集成

expand[ɪkˈspænd] *v.* 扩展

strengthen[ˈstreŋθən] *v.* 加强;强化

foundation[faʊnˈdeɪʃ(ə)n] *n.* 基础

hierarchy[ˈhaɪərɑːki] *n.* 层级

seamlessly[ˈsiːmləsli] *adv.* 无缝地

comprehensive[ˌkɒmprɪˈhensɪv] *adj.* 全面的

challenge[ˈtʃælɪndʒ] *n.* 挑战

assume[əˈsjuːm] *v.* 假设

modify[ˈmɒdɪfaɪ] *v.* 修改

interconnectivity[ɪntəkənekˈtɪvɪti] *n.* 互联互通

silos[ˈsaɪləʊz] *n.* 孤岛

varying[ˈveəriɪŋ] *adj.* 不同的

ongoing[ˈɒngəʊɪŋ] *adj.* 持续的

vendor[ˈvɛndə] *n.* 供应商

subcontractor[ˌsʌbkənˈtræktə] *n.* 分包商

agility[əˈdʒɪlɪti] *n.* 敏捷性

flexibility[ˌflɛksɪˈbɪlɪti] *n.* 灵活性

transparency[trænsˈpærənsi] *n.* 透明度

be used for 用于

lack of 缺乏

break down 打破,分解

be equipped with 配备有

in other words 换句话说

pay off 取得成功,得到回报

result in 导致

Technical Terms

horizontal integration 横向集成

vertical integration 纵向集成

engineering process 工程流程

knowledge silos 知识孤岛

asset management 资产管理

supply chain management 供应链管理

data security 数据安全

Task 4 *According to the passage, are the following statements true (T) or false (F)?*

(　　) 1. Horizontal integration in manufacturing involves connecting machines, equipment, and production units.

(　　) 2. Vertical integration in manufacturing only involves connecting sensors to the business level.

(　　) 3. Horizontal integration allows the machinery and IoT devices to work together with no need for seamless engineering processes.

(　　) 4. Vertical integration enables production data to be used for business decisions by facilitating communication with ERP systems.

(　　) 5. End-to-end integration can be achieved by combining horizontal integration and vertical integration.

(　　) 6. One of the challenges of horizontal and vertical integration is lack of communication, which may need the purchase of new IoT devices.

(　　) 7. Knowledge silos are primarily a challenge in vertical integration but not in horizontal integration.

(　　) 8. Horizontal integration naturally leads to better data security and privacy practices.

Task 5 *Find the answers to the following questions in the passage.*

1. What is horizontal integration in manufacturing?

____

2. What is vertical integration in manufacturing?

____

3. What is the significance of combining horizontal integration and vertical integration in a business?

____

4. Why is breaking down knowledge silos considered as a challenging task?

____

5. What is the challenge of data sharing in horizontal integration?

____

Task 6 *Match the English expressions in Column A with their Chinese meanings in Column B.*

Column A	Column B
1. horizontal integration	A. 通信协议
2. vertical integration	B. 供应链管理
3. engineering process	C. 纵向集成
4. Communication Protocol	D. 资产管理
5. end-to-end integration	E. 横向集成
6. supply chain management	F. 端到端集成
7. asset management	G. 数据孤岛
8. data silos	H. 工程流程

Task 7 *Make six sentences with words from the following boxes.*

horizontal | vertical | seamless | comprehensive

real-time | data-driven | varying | cost-effective

Example: The integration of IoT devices allows for **real-time** communication between machines, enhancing the precision and speed of the manufacturing process.

1. ____.
2. ____.
3. ____.
4. ____.
5. ____.
6. ____.

Task 8 *Translate the following sentences into English using the words and expressions given.*

1. 睡眠不足会导致很多健康问题。(lack of)

____.

2. 成熟的人通常能应对生活中的巨大压力。(be capable of)

____.

3. 宇航员们配备了特殊服装,以保护他们免受太空恶劣条件的影响。(be equipped with)

____.

4. 众所周知,失败通常源于懒惰,而勤奋可以带来成功。(result in)

____.

5. 你的努力总有一天会得到回报的。(pay off)

____.

Investigating

▶ Horizontal integration and vertical integration may seem challenging in intelligent manufacturing, but the industrial internet platform (IIP) can help overcome these integration difficulties. In this section, we will explore these platforms.

在智能制造中达成横向集成和纵向集成可能有困难,但工业互联网平台可以帮助解决这些集成难题。在本部分,我们将探讨什么是工业互联网平台。

Industrial Internet Platform

The Industrial Internet Platform has introduced a new development model for the transformation and upgrading of the global manufacturing industry. It presents a significant opportunity for the leapfrog development of China's manufacturing industry.

Understanding the Industrial Internet Platform

The Industrial Internet encompasses four main systems: network, platform, data, and security. It serves as the infrastructure for the digital, networked, and intelligent transformation of industries, integrating the Internet, big data, and AI with the real economy. The Industrial

Internet Platform acts as the "operating system" of the Industrial Internet, including edge, infrastructure as a service (IaaS), platform as a service (PaaS), and software as a service (SaaS) layers. It has four main functions.

- Data aggregation: Collects diverse, massive data at the network level for deep analysis and application.
- Modeling and analysis: Provides big data and AI analysis models and simulation tools to achieve data-driven decisions and intelligent applications.
- Knowledge reuse: Converts industrial expertise into model libraries and knowledge bases for easy redevelopment and widespread use.
- Application innovation: Offers industrial applications and cloud software to enhance efficiency in research and development (R&D), equipment management, enterprise operations, and resource scheduling.

Through these functions, the Industrial Internet Platform enhances production efficiency and management, driving digital and intelligent transformation across industries.

RootCloud Industrial Internet Platform

RootCloud IIP is a leading provider of intelligent manufacturing and Industrial Internet of Things (IIoT) solutions. It has introduced a new development model for the transformation and upgrading of the global manufacturing industry. The platform is widely applied across various levels and links of the industrial system.

1. Horizontal Integration with RootCloud

In the typical manufacturing environment, horizontal integration must connect many types of equipment. RootCloud excels in horizontal integration by seamlessly connecting various industrial devices and systems, supported by several key features.

- Compatibility with over 1,100 industrial protocols: Ensures smooth communication between disparate machines.
- Real-time data collection: Facilitates immediate analysis and action across connected devices.
- Edge computing: Allows for data acquisition, processing, and analysis in disconnected scenarios.
- Unlimited connectivity: Supports over 22 industries and nearly 1.8 million industrial devices.

2. Vertical Integration with RootCloud

Vertical integration creates connections between production and other parts of a manufacturing organization, networking beyond traditional production hierarchy levels—from the sensor to the business level. RootCloud is ideal for vertical integration, offering essential features.

- ERP interface: Integrates with enterprise resource planning systems like SAP[①] and Microsoft Dynamics.
- Comprehensive lifecycle management: Supports the entire lifecycle of products, from design to maintenance.
- Cloud integration: Facilitates data storage, processing, and analysis in the cloud for enhanced accessibility and scalability.
- AI and big data: Use AI and big data to provide insights and drive intelligent decision-making.

3. Overcoming Integration Challenges with RootCloud

Horizontal integration and vertical integration may seem challenging in a smart factory, but RootCloud helps overcome these difficulties with its robust, hardware-independent platform. Key benefits include the followings.

- Flexible and scalable solutions: Integrate with existing systems and scale to accommodate future growth.
- User-friendly interface: Provides an intuitive graphical interface for easy management and monitoring.
- Security and compliance: Ensures secure data handling and compliance with local and international data privacy laws.

RootCloud has helped numerous companies, such as SANY[②], achieve significant improvements in efficiency and productivity, solidifying its position as a leader in industrial digitalization.

New Words and Phrases

transformation [ˌtrænsfəˈmeɪʃn] *n.* 转型

① SAP(systems, applications & products in data processing,中文名为恩爱普)于1972年成立于德国,是全球领先的企业应用软件提供商。

② SANY,指三一集团,是一家全球知名的工程机械制造商,成立于1989年。

upgrading[ˌʌpˈgreɪdɪŋ] *n.* 升级

aggregation[ˌægrɪˈgeɪʃn] *n.* 聚合

massive[ˈmæsɪv] *adj.* 大量的

modeling[ˈmɒd(ə)lɪŋ] *n.* 建模

simulation[ˌsɪmjʊˈleɪʃ(ə)n] *n.* 仿真

reuse[ˌriːˈjuːz] *n.* 再使用;重复使用

redevelopment[ˌriːdɪˈveləpmənt] *n.* 重新开发

compatibility[kəmˌpætəˈbɪlɪti] *n.* 兼容性

disparate[ˈdɪspərət] *adj.* 不同的

acquisition[ˌækwɪˈzɪʃ(ə)n] *n.* 获得

interface[ˈɪntəfeɪs] *n.* 接口

accommodate[əˈkɒmədeɪt] *v.* 适应

compliance[kəmˈplaɪəns] *n.* 合规

solidify[səˈlɪdɪfaɪ] *v.* 巩固

Technical Terms

Industrial Internet Platform(IIP)　工业互联网平台

edge layer　边缘层

infrastructure as a service(IaaS)　基础设施即服务

platform as a service(PaaS)　平台即服务

software as a service(SaaS)　软件即服务

data aggregation　数据聚合

modeling and analysis　建模和分析

RootCloud Industrial Internet Platform(IIP)　根云工业互联网平台

cloud integration　云集成

Task 9　*Read the passage and match the two parts of the sentences.*

1. The Industrial Internet encompasses	A. data acquisition, processing, and analysis in disconnected scenarios.
2. The Industrial Internet Platform acts as	B. four main systems: network, platform, data, and security.

Continued

3. The Industrial Internet Platform collects	C. over 22 industries and nearly 1.8 million industrial devices.
4. The Industrial Internet Platform converts	D. the "operating system" of the Industrial Internet, including edge, IaaS, PaaS, and SaaS layers.
5. RootCloud IIP is	E. an intuitive graphical interface for easy management and monitoring.
6. RootCloud edge allows for	F. diverse, massive data at the network level for deep analysis and application.
7. RootCloud IIP supports	G. a leading provider of intelligent manufacturing and IIoT solutions, widely applied across various levels and links of the industrial system.
8. RootCloud IIP provides	H. industrial expertise into model libraries and knowledge bases for easy redevelopment and widespread use.

Task 10 *Translate the sentences that you completed in Task 9. The first one is provided.*

1. The Industrial Internet encompasses four main systems: network, platform, data, and security.
工业互联网包含了网络、平台、数据、安全四大体系。

2.

3.

4.

5.

6.

7.

8.

Task 11 *Listen to a recording about MES. Fill in the blanks with the words provided in the box and read it aloud.*

• downtime	• finished	• real-time	• life-cycle
• services	• computerized	• tracking	• scheduling

MES are 1. ________ systems used in factories to track and record how raw materials become 2. ________ products. MES gives important information to help managers improve production. It monitors the production process in 3. ________, controlling things like materials, workers, machines, and support 4. ________ .

MES can help in many areas, such as managing product details across the product 5. ________, 6. ________ resources, executing and dispatching orders, analyzing production, managing 7. ________ to improve equipment efficiency, ensuring product quality, and 8. ________ materials.

Researching

▶ How do Chinese lighthouse factories light up the journey to new industrialization? Let's study and analyze some cases together.

中国的灯塔工厂如何引领新型工业化？让我们一起研究和分析一些案例。

灯塔工厂是麦肯锡(McKinsey)和世界经济论坛(World Economic Forum)在2019年提出的全新概念,其代表了当今全球制造业智能制造和数字化的最高水平。目前,全球有153家灯塔工厂,其中有62家在中国,工厂数量位居全球首位。我国的灯塔工厂正在不断丰富“智造”内涵,并努力将其成功经验复制、推广。

Task 12 *Case Study: China's Lighthouse Factories* (*Fig. 4.2*)

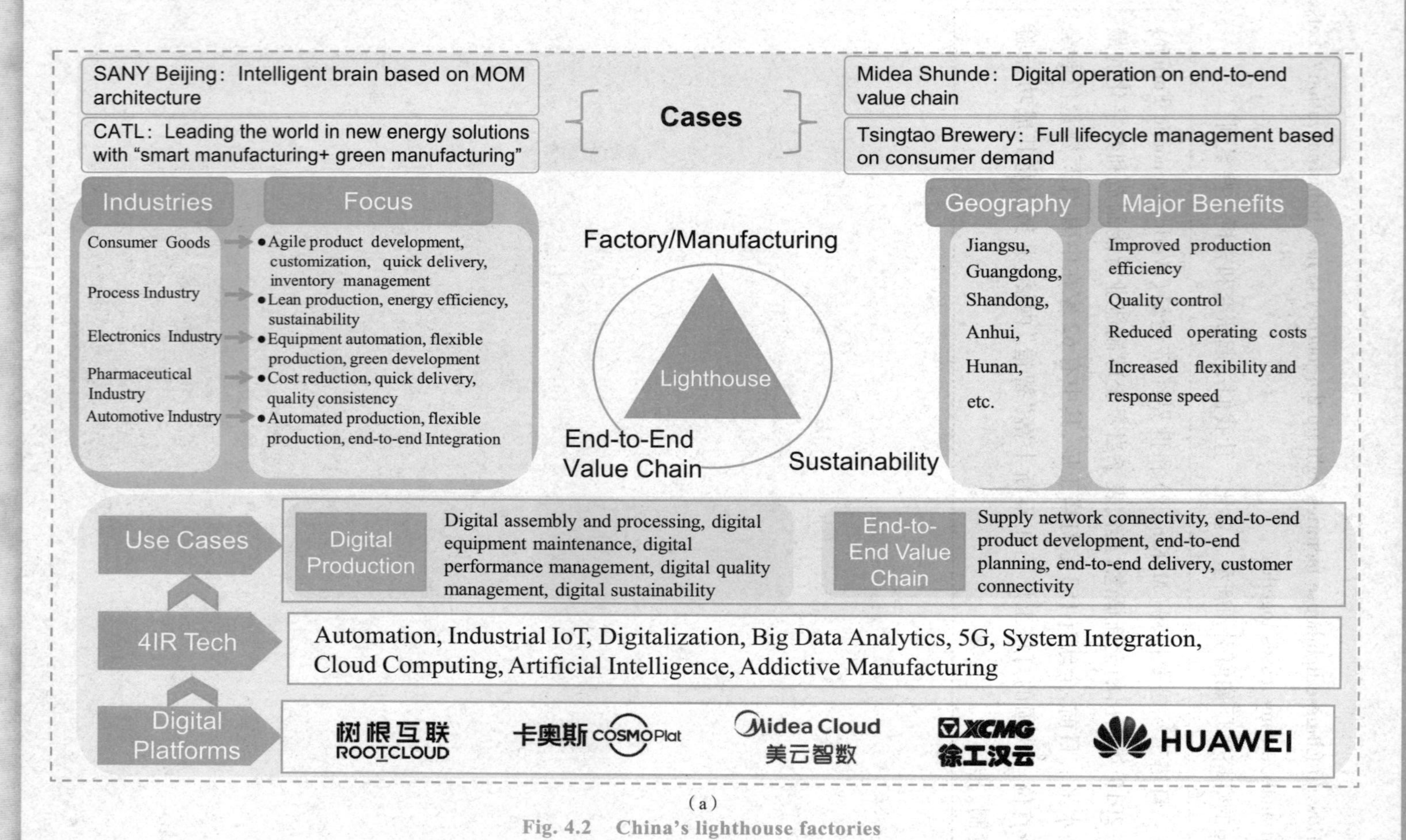

(a)

Fig. 4.2 China's lighthouse factories

(a) English version

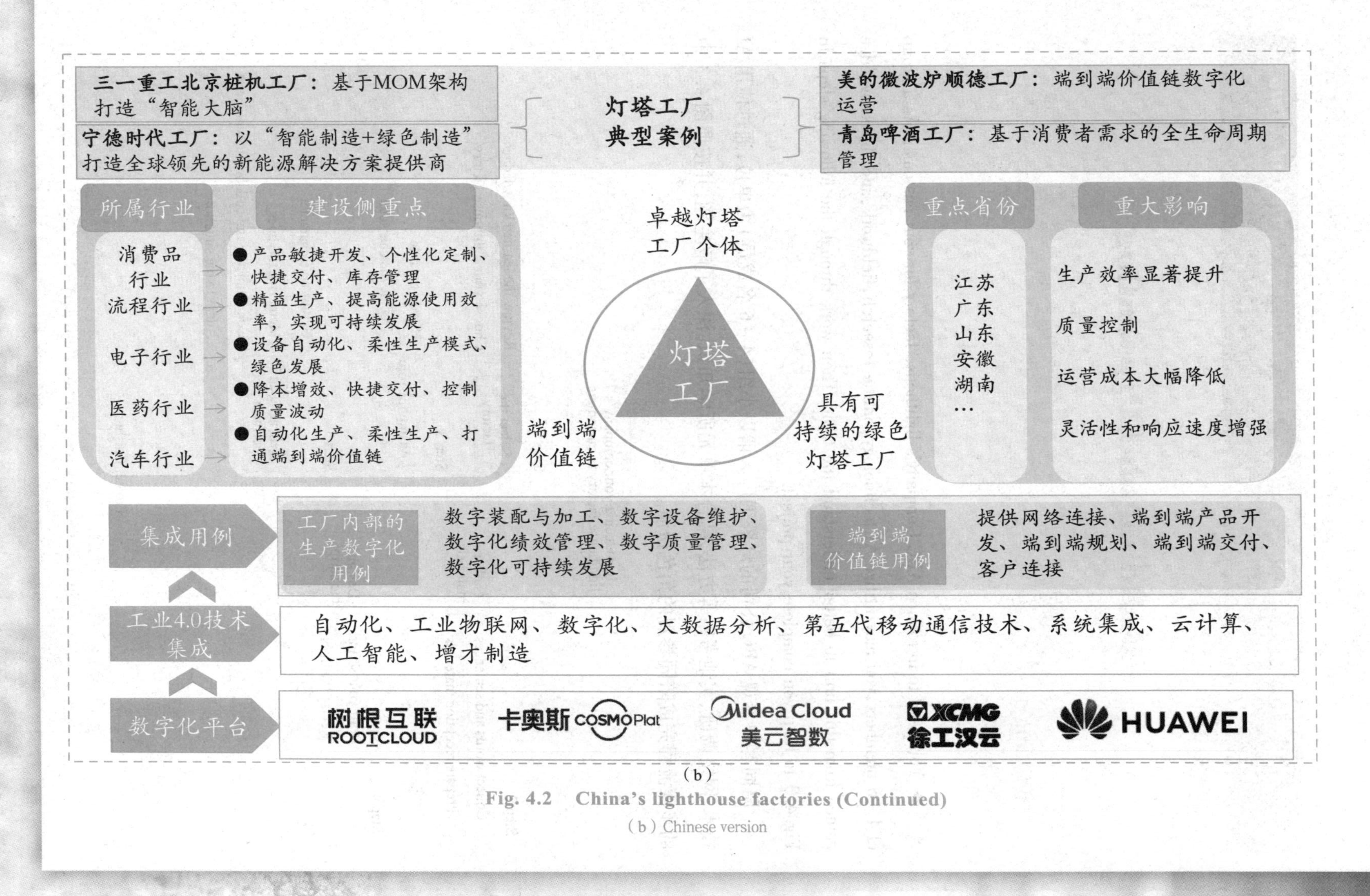

(b)

Fig. 4.2 China's lighthouse factories (Continued)

(b) Chinese version

Task 13 *Cases Analysis: Analyze the changing stories of Chinese lighthouses like SANY.*

Building and Presenting

Building

实施

▶ You need to introduce ABC Company's lighthouse factory for your customer. As a group of 4 – 6 members, you may utilize AI technology, online research, fieldwork, and the knowledge gained from this unit to better understand the topic. Then work through the following steps in Fig. 4. 3 to help you complete your project.

陪同客户参观 ABC 公司的灯塔工厂。请以小组(4 ~ 6 名成员)为单位,通过使用 AI 技术、网络调研、实地考察等方法,结合本单元所学知识,深入了解主题,并按照图 4. 3 中的步骤流程完成陪同参观的角色扮演。

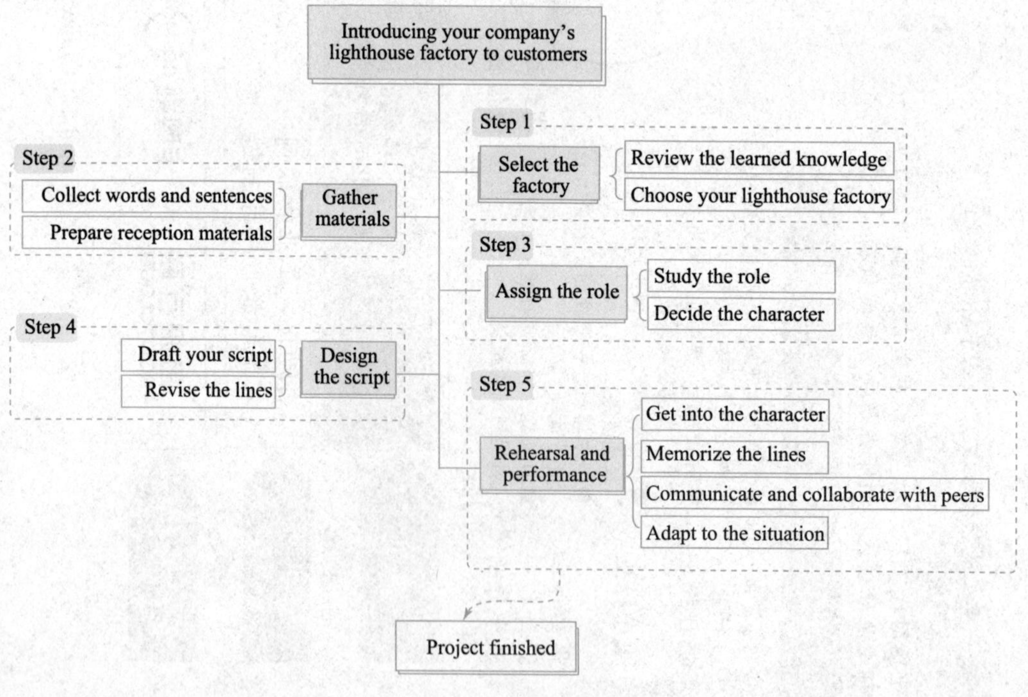

Fig. 4. 3 Factory introduction flowchart

Task 14

Step 1 Select the Factory

A. Review the learned knowledge

Answer the following questions in Table 4. 1 based on what you've learned before.

Table 4. 1 List of questions under three different topics

Topics	Questions	Notes
Integration	What are horizontal integration and vertical integration in manufacturing?	
	What are the challenges of horizontal integration and vertical integration?	
Industrial Internet Platform	What is Industrial Internet Platform?	
	What are the key benefits of RootCloud Industrial Internet Platform?	
Lighthouse factories	What lighthouse factories do you know?	
	What technologies are used in lighthouse factories?	

B. Choose your lighthouse factory

Search for information about the following four companies and their lighthouse factories in Table 4. 2, select one of them and complete the content in the table. Or you can pick another one as you like.

Table 4. 2 The companies and their lighthouse factories

Companies	Lighthouses	Industries	Use Cases	Major Benefits
SANY				
Midea				
Lenovo				
GAC AION				
Others				

Task 15

Step 2 Gather Materials

A. Collect words and sentences

Brainstorm words or expressions related to integration and lighthouse factories (Fig. 4. 4), and then make sentences with them.

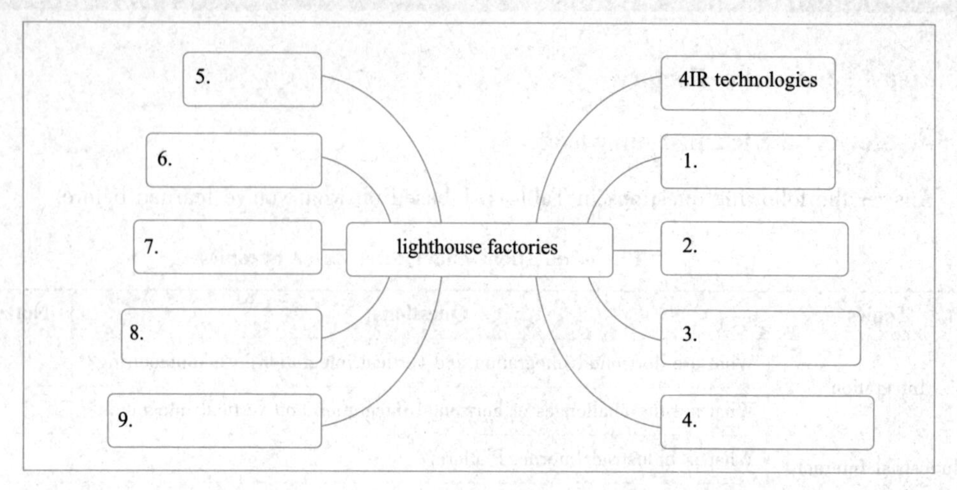

Fig. 4.4 Words and expressions related to lighthouse factories

Example: Midea leverages 4IR technologies to transform from an automated factory to an end-to-end connected value chain, improving labour efficiency by 28%, reducing unit cost by 14% and shortening order lead time by 56%.

1.

2.

3.

4.

5.

6.

B. Prepare reception materials

Prepare reception materials, such as brochures, safety instructions and factory maps.

Task 16

Step 3 Assign the Role

A. Study the role

Study the role and understand the duties involved during the reception (Table 4.3).

Table 4. 3 Roles and duties

Roles	Duties	Name
Client	You are interested in the operations and products of the factory.	
Receptionist	You are responsible for greeting clients, introducing basic information about the company and the lighthouse factory.	
Factory guide	You need to provide detailed explanations to various parts of the factory.	
Technical expert	You need to answer the client's questions on technical details.	
Sales representative	You need to discuss product sales, share latest products.	

2. Decide the character

Decide on a name based on what you know about your character, and fill in Table 4. 3.

Task 17

Step 4 Design the Script

A. Draft your script

Cooperate with your group members and draft the script for different scenarios according to Table 4. 4, making sure it covers the key points in Task 14.

Table 4. 4 Scenarios and key points

Scenarios	Key Points	Lines
Client arrival	Greeting and self-introduction	
Company introduction	• Introduce the company overview, the characteristics of the lighthouse factory, and main products	
Safety instructions	• Explain the safety rules and precautions	
Factory tour	• Demonstrate the lighthouse factory's achievements in automation, digitalization, and intellectualization • Explain the technologies and applications	
Equipment demonstration	• Explain its features, advantages, and operational processes	
Interactive experience	• Experience the operation of equipment • Answer clients' questions	
Feedback collection	• Gather feedback from clients	
Follow-up	• Share updates on the company's latest developments, technological advancements • Discuss cooperation opportunities	

B. Revise the lines

Polish your lines according to language skills you learned in Task 2(Table 4.5).

Table 4.5 Comparison of wording in the lines and wording after polishing

Wording in the lines	Wording after polishing

Task 18

Step 5 Rehearsal and Performance

After finishing your script, refer to the following advice in Fig. 4.5, and mark each one as accomplished as you go until you finish your role-play.

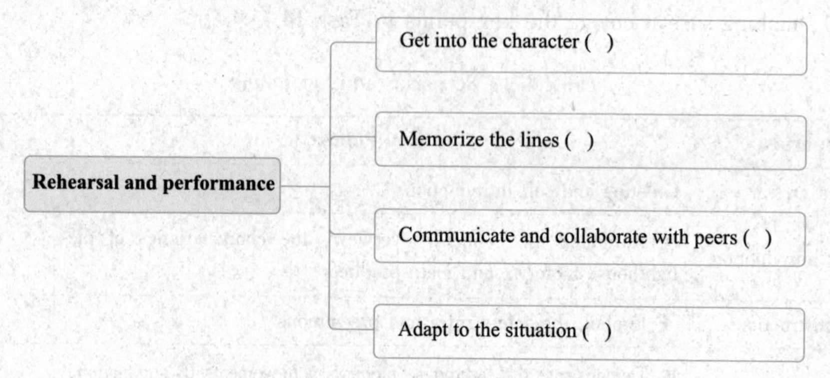

Fig. 4.5 Rehearsal and performance advice

Presenting
展示

▶ Now that you have finished your rehearsal and preparations, it is time to act it with your group members in class.

排练及准备工作都完成后,即请与同学们在课上分享你们组的演出。

Task 19 *Take questions, comments, or suggestions from the audienceif possible (Table 4.6).*

Table 4.6 Feedback from the teacher and classmates

Feedback from the teacher	
Feedback from classmates	

Task 20 *Modify the script and record your performance after class. The group leader sends the videos to the teacher's mailbox or uploads them to the online platform as the usual performance evaluation document and for everyone to learn from online.*

Assessing and Reflecting

Assessing
评估

▶ Assessing the work you have accomplished allows you to know not only the response of your audience to your work but also the weak parts of your work to improve.

评估你所完成的工作,不仅能了解观众对你作品的反馈,还能发现不足,以便改进。

Task 21 *The group project is evaluated jointly by teachers and students (Table 4.7).*

Table 4.7 Group project evaluation form

Evaluation (*Excellent* =3, *Good* =2, *Poor* =1)					
Group	**Language**	**Creativity**	**Presentation**	**Collaboration**	**Cross-subject(role-play)**
Group1					
Group2					
Group3					
Group4					
Group5					

Reflecting

反思

▶ Reflecting on your learning helps you understand what you've gained from this unit and how you can apply it in the future.

反思有助于你理解在本单元中所学的知识,以及将来如何加以运用。

Task 22 *The following questions in Table 4.8 may help you do the reflection, but feel free to ask yourself more questions when necessary.*

Table 4.8 Questions reflection form

Questions	Reflections
Did I increase my vocabulary to finish my task more precisely?	
Did I use any digital technologies?	
Did I actively engage in preparing the project?	
Did I develop my performing skills through role-play?	
Did my contributions help others understand integration in IM better?	
Did I identify, analyze and solve problems by fulfilling various tasks?	

Extending

▶ You are a technical engineer at a new energy manufacturing company. Faced with industry challenges, you visit CATL to investigate how a battery is produced and how the factory earned its status as a lighthouse factory. After the visit, please write a report focusing on the factory's smart manufacturing and green manufacturing initiatives.

你是某新能源制造企业的技术工程师,面对行业挑战,你前往宁德时代参观调研,了解一块电池如何诞生、一座灯塔如何点亮。参观结束后,请撰写一份报告,重点说明工厂智能制造、绿色制造的情况。

Task 1 *Get acquainted with CATL Lighthouses.*

On December 14, 2023, CATL's Liyang plant was officially recognized as a "Lighthouse Factory" by the World Economic Forum (WEF). This achievement marks CATL's third lighthouse factory, joining its production bases in Ningde and Yibin. Currently, all three Lighthouses in the global lithium battery industry are from CATL.

图片来源:先进制造 AMC 公众号:走进灯塔工厂(12):宁德时代

Task 2 *Familiarize yourself with the typical technical solutions of CATL Lighthouses.*

Core Principles	Technical Solutions	Description
High Automation	Automated production lines	*Streamline all stages of battery production, including assembly, testing and packaging.*
	Intelligent management systems	*Enable real-time monitoring and efficient task scheduling for smooth production.*
	3D printing technology	*Reduces line-switching time and enhances manufacturing flexibility.*

Continued

Core Principles	Technical Solutions	Description
High Automation	Computer vision technology	*Enables quality inspections at the micrometer level to ensure precision and consistency.*
	Big data analysis	*Optimizes workflows, improves product quality and enhances productivity.*
	5G technology	*Provides full network coverage and supports new-zero defect manufacturing in the factory.*
	Digital technology	*Reduces costs, improves delivery capability, and reshapes labor-production dynamics.*
Smart Management	Digital management	*Combines IoT, AI, and cloud computing to monitor data in real time, ensuring supply chain transparency and traceability.*
	AI inspection technology	*Analyzes images in real time using algorithms, quickly identifying and marking defective parts on each production line.*
	Highly automated processes	*Reduce human intervention in cell manufacturing and assembly.*
	Sensors and AI systems	*Over 40,000 sensors monitor plant conditions, while* AI systems *optimize energy-saving strategies.*
Green Manufacturing	Energy management optimization	*Uses advanced algorithms to optimize production and improve accuracy.*
	Clean energy integration	*Utilizes solar power system to enhance renewable energy use.*
	Digital energy-saving systems	*Employs automated monitoring tools to streamline operations and boost efficiency.*
	Zero-carbon certification	*Achieved recognition as a net-zero emissions factory in January* 2023.
	Green supply chain management	*Promotes sustainable practices across industries, reducing waste and emissions in operations.*

Task 3 *Complete the following report based on Task 1 and Task 2.*

CATL Lighthouse Factory Report			
Researcher	Engineer Zhao	Date	May 15, 2024
Background	CATL's three lighthouse factories have set benchmarks with their adoption of digital technologies and green manufacturing. This research aims to analyze how these technologies are applied upstream and downstream.		

Continued

CATL Lighthouse Factory Report		
Scope	Automated production lines Smart management platforms Green manufacturing zones	
Methods	Factory visits, in-depth interviews, face-to-face expert discussions	
Core Principles	**Workplace Scenarios**	**Technical Solutions**
High automation	To enhance flexibility and reduce line-switching time, which solution should be chosen?	1. ________
	To reduce cost and improve delivery capacity, which solution should be chosen?	2. ________
	For micrometer-level quality inspections, which solution should be chosen?	3. ________
Smart management	For real-time production data monitoring and supply chain transparency, which solution should be chosen?	4. ________
	To quickly identify and mark defective parts, which solution should be chosen?	5. ________
	To optimize energy-saving strategies, which solution should be chosen?	6. ________
Green manufacturing	To reduce energy consumption and improve efficiency, which solution should be chosen?	7. ________
	To increase renewable energy usage, which solution should be chosen?	8. ________
	To minimize energy consumption and carbon emissions across the supply chain, which solution should be chosen?	9. ________

Unit 5
Innovation in Intelligent Manufacturing
智能制造中的创新

Unit Introduction

Driving Question

驱动问题

As the manufacturing industry faces rapid technological changes, embracing innovation is no longer optional but essential. Emerging technologies such as AI and digital twins are not only innovations themselves but also key drivers behind the innovation of manufacturing. How do AI and these technologies lead to breakthroughs and advancements in the manufacturing sector?

鉴于制造业所面临的快速的技术变革,追求创新已成为当务之急。人工智能和数字孪生等技术不仅是技术自身的创新,也是推动制造业创新的关键力量。那么,AI 及相关技术如何助力制造业创新呢?

Project Overview

项目概述

In this unit, you will begin by exploring the basic concepts of innovation in manufacturing. After that, you will delve into the innovative applications of AI in the manufacturing industry. Next, you will learn about the concept of digital twins. Then, you will study practical application cases of how AI is being used to innovate within China's manufacturing industry. Next, working in groups, you will prepare a PPT presentation showcasing ABC Company's innovative applications of AI in manufacturing. After the presentation, you will discuss the benefits and challenges of using AI and related technologies to foster innovation in manufacturing.

在本单元,首先学习制造业中有关创新的基本概念,然后深入学习人工智能在制造业中的创新应用,了解数字孪生的概念,研究人工智能在中国制造业中实际应用于创新的案例。随后,小组合作制作一份 ABC 公司如何创新应用 AI 的 PPT 演示文稿并进行汇报。最后,讨论在制造业中使用 AI 及相关技术进行创新的优势和挑战。

Project Extension

项目拓展

After completing the main project, you'll engage in a hands-on extension activity to deepen your understanding of digital twin technology. This will involve learning how to complete virtual-real interaction on the production line.

在完成主要项目后，将进行实操拓展，实现虚实联动，以进一步加深对数字孪生等技术的理解。

Learning Objectives

学习目标

This unit is intended to help you:

1. have a general idea of innovation in manufacturing;
2. explore the applications of AI in manufacturing;
3. analyze case studies of AI applications in manufacturing industry in simple English;
4. prepare an English presentation showcasing AI applications in manufacturing;
5. complete virtual-real interaction on the production line;
6. cultivate innovative thinking.

本单元旨在帮助你：

1. 了解制造业中的创新；
2. 探讨人工智能技术在制造业中的创新应用；
3. 用英语简单分析人工智能在制造业中的创新应用案例；
4. 做英文 PPT 演示，展示人工智能在制造业中的应用；
5. 复现真实生产线如何实现虚实联动；
6. 培养创新思维。

Getting Ready

► What is a PPT presentation? What are the commonly used language for a presentation? Please complete the tasks below to find the answer. Fig. 5.1 shows a man giving a PPT presentation.

什么是 PPT 演示？进行 PPT 演示时有哪些常用表达？请完成以下任务，找到答案。在图 5.1 中，一位男士正在进行 PPT 演示。

Fig. 5. 1 PPT presentation

Task 1 *Listen to a recording about a product presentation and fill in the blanks with the missing words.*

A good PPT presentation is akin to a well-crafted 1. ________, and like any compelling narration, it is structured into three 2. ________ parts: introduction, body, and conclusion. The introduction sets the tone for the entire presentation and explains what the audience will come away with after viewing it. Following this, the body emerges as the 3. ________ component of your presentation, tasked with upholding the 4. ________ made during the introduction. It is within this section that you delve into the details of your 5. ________ and systematically present all your information. Depending on the complexity and nature of your presentation, it is advisable to break it down into manageable segments or key points. These points should be organized in a 6. ________ sequence, with each one supported by relevant and 7. ________ information. Finally, a good conclusion summarizes the key points you made or highlights what the audience should have learned. It clarifies the general 8. ________ of your presentation and reinforces the reason for viewing it.

Task 2 *Listen to the following presentation and tick the phrases you hear.*

( )1. Welcome to...

( )2. Thank you for joining us here at...

( )3. What I'd like to talk to you about today is...

( )4. Today's focus will be on...

( )5. You're in the right place.

( )6. You've come to the perfect spot...

( )7. Let me walk you through...

( )8. Let me give you an idea of what I will be talking about.

( )9. It's clear to all of us that...

( )10. I think we all agree that...

( )11. To begin with, I'd like to emphasize...

( )12. First of all, I'd like to highlight...

( )13. What I was saying was that...

( )14. What I mean was...

( )15. To sum up,...

( )16. To put in a nutshell,...

( )17. I'd love to hear any questions you may have about our product.

( )18. I'd be happy to answer any questions you have about our product.

( )19. Fire away.

( )20. Go ahead.

Task 3 ***Sort the phrases from Task 2 into the mind map, as shown in Fig. 5. 2.***

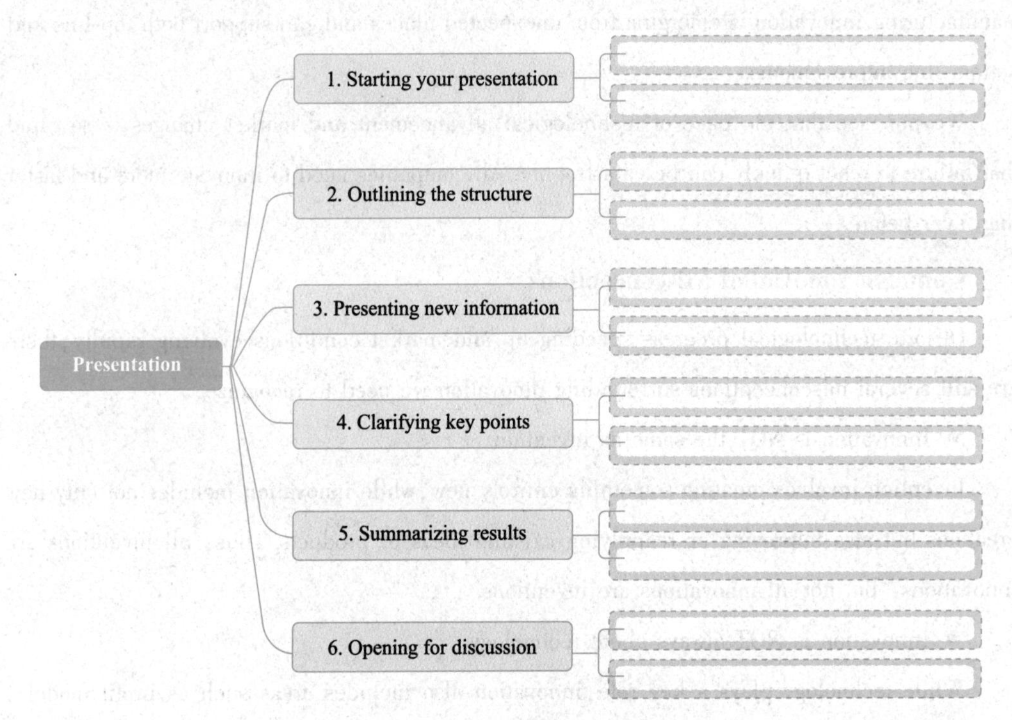

Fig. 5. 2 Mind map for giving a presentation

Starting up

▶ What does true innovation mean in today's rapidly changing world? Are there misconceptions about innovation in manufacturing that might be holding you back? In this section, we will learn some concepts of innovation in manufacturing.

在当今快速变化的世界中,真正的创新意味着什么？在制造业中,是否有一些对创新的误解阻碍了我们的发展？在本部分,我们将会学习一些制造业中的创新概念。

Innovation in Manufacturing

Technology is advancing at an unprecedented pace, driving the need for innovation across all industries, including manufacturing today.

An Introduction about Innovation

Innovation is the practical application of ideas that result in the introduction of new goods and services or improvements in the way goods and services are offered. With that in mind, manufacturing innovation is emerging from unexpected places and can support both top-line and bottom-line improvements.

We now see that the pace of technological advancement and market changes is so rapid that failure to react quickly can be catastrophic. All companies need to innovate more and faster than ever before.

Common Innovation Misconceptions

Despite technological progress speeding up and market conditions evolving rapidly, there are still several misconceptions surrounding innovation we need to recognize.

- Innovation is NOT the same as invention.

Invention involves creating something entirely new, while innovation includes not only new creations but also improving or reapplying existing ideas or products. Thus, all inventions are innovations, but not all innovations are inventions.

- Innovation is NOT always about technology.

While technology plays a key role, innovation also includes areas such as profit models, networks, services, channels, structures, processes, product performance, and brand engagement. Most major breakthroughs combine several types of innovation.

- Innovation doesn't occur in sudden moments of inspiration.

It is rarely a sudden flash of genius. Instead, innovation is a prolonged process involving experimentation, failure, learning, and refinement.

- You don't have to be inherently creative to be innovative.

Innovation is a skill that anyone can learn and practice. Often, the best innovations come from diverse people with different perspectives.

Formalizing Innovation for Manufacturing

Making innovation formal helps align creative efforts with strategic goals, creating a structured process for generating and executing ideas. It ensures resources are allocated and processes are set up to support, test, and implement new ideas.

Here are some steps to formalize innovation.

• Create an innovation strategy.

Align innovation efforts with the company's strategic objectives. Determine the focus areas (product, process, business model) and set goals and metrics for success.

• Establish an innovation leader and team.

Appoint individuals or a team responsible for leading, managing, and tracking innovation efforts.

• Cultivate an innovative culture.

Encourage creative thinking, risk-taking, and challenging the status quo. Foster open communication, collaboration, and recognize and reward innovative ideas.

• Implement an idea management system.

Develop a system to collect, evaluate, and implement new ideas, ranging from simple suggestion boxes to digital platforms for idea submission and collaboration.

As the manufacturing industry enters this new era marked by technological disruption, it is crucial to embrace innovation. By adapting to changing technology and formalizing innovation processes, we can confidently navigate this transformative period, stay competitive, and drive growth in a rapidly evolving world.

New Words and Phrases

unprecedented[ʌnˈpresɪdentɪd] *adj.* 前所未有的

improvement[ɪmˈpruːvmənt] *n.* 改进,提升

emerge[ɪˈmɜːdʒ] *v.* 出现

advancement[ədˈvænsmənt] *n.* 进步

catastrophic[ˌkætəˈstrɒfɪk] *adj.* 灾难性的

innovate[ˈɪnəveɪt] *v.* 创新

evolve[ɪˈvɒlv] *v.* 进化,发展

misconception[ˌmɪskənˈsepʃ(ə)n] *n.* 误解

recognize[ˈrekəgnaɪz] *v.* 认识

invention[ɪnˈvenʃ(ə)n] *n.* 发明

entirely[ɪnˈtaɪəli] *adv.* 完全地

reapply[ˌriːəˈplaɪ] *v.* 重新应用

channel[ˈtʃænl] *n.* 渠道

breakthrough[ˈbreɪkθruː] *n.* 突破

inspiration[ˌɪnspɪˈreɪʃ(ə)n] *n.* 灵感

rarely[ˈreəli] *adv.* 很少地

flash[flæʃ] *n.* 闪现

prolonged[prəˈlɒŋd] *adj.* 持久的

experimentation[ɪkˌsperɪmenˈteɪʃn] *n.* 实验

refinement[rɪˈfaɪnmənt] *n.* 改进

inherently[ɪnˈhɪərəntli] *adv.* 天生地

align[əˈlaɪn] *v.* 对齐

execute[ˈeksɪkjuːt] *v.* 执行,实施

allocate[ˈæləkeɪt] *v.* 分配

formalize[ˈfɔːməlaɪz] *v.* 使形式化

appoint[əˈpɔɪnt] *v.* 任命,指派

cultivate[ˈkʌltɪveɪt] *v.* 培养

foster[ˈfɒstə(r)] *v.* 促进,培养

evaluate[ɪˈvæljueɪt] *v.* 评估

submission[səbˈmɪʃ(ə)n] *n.* 提交

embrace[ɪmˈbreɪs] *v.* 拥抱

navigate[ˈnævɪgeɪt] *v.* 应对,驾驭

emerge from　从……中浮现;幸存下来

speed up　加速

the same as　与……一样

set up　建立

range from... to　范围从……到

Technical Terms

technological advancement　技术进步

profit models　利润模型

product performance　产品性能

brand engagement　品牌参与度

Task 4 *According to the passage, are the following statements true (T) or false (F)?*

(　　)1. Technology is advancing at a rapid pace, driving the need for manufacturing innovation.

(　　)2. Innovation means creating entirely new products or services.

(　　)3. All inventions are considered innovations, but not all innovations are inventions.

(　　)4. Innovation only focuses on technological advancement.

(　　)5. Innovation is a sudden moment of inspiration or a flash of genius.

(　　)6. An innovation strategy should be aligned with the company's strategic objectives and focus areas.

(　　)7. It is encouraged to develop a system to collect, evaluate, and implement new ideas.

(　　)8. Cultivating an innovative culture does not involve recognizing and rewarding innovative ideas.

Task 5 *Find the answers to the following questions in the passage.*

1. What does "innovation" refer to according to the text?

____

2. How does the text describe the process of innovation?

____

3. Is innovation the same as invention? How to distinguish them?

____

4. What steps can companies take to formalize innovation?

____

5. Why is it important to formalize innovation in manufacturing?

____

Task 6 *Match the English expressions in Column A with their Chinese meanings in Column B.*

Column A	Column B
1. technological advancement	A. 技术进步
2. profit models	B. 创意管理系统
3. product performance	C. 品牌参与度
4. brand engagement	D. 技术颠覆
5. innovative strategy	E. 产品性能
6. innovative culture	F. 创新文化
7. idea management system	G. 创新策略
8. technological disruption	H. 利润模型

Task 7 *Make six sentences with words from the following boxes.*

improvement　advancement　innovation　invention

misconception　inspiration　experimentation　submission

Example: Continuous **improvement** in manufacturing processes has helped the company maintain its competitive edge in the market.

1. ____.
2. ____.
3. ____.
4. ____.
5. ____.
6. ____.

Task 8 *Translate the following sentences into English using the words and expressions given.*

1. 为了加快交付流程和提高客户满意度,公司开始使用新技术。(speed up)

____.

2. 找到一份符合自身价值观的工作,更容易带给你满足感。(align... with)

____.

3. 那三个研究生曾想创办自己的出口公司。(set up)

____.

4. 高压会引起疲劳和紧张性头痛。(emerge from)

____.

5. 工厂利用智能系统来分配机器和工人的工作,这样可以提高生产效率,避免资源浪费。(allocate)

____.

Investigating

▶ AI technologies are transforming traditional manufacturing processes by enabling autonomous tasks, improving accuracy, and enhancing maintenance and design. In this section, we'll explore AI applications in the manufacturing industry.

通过实现自主任务、提高准确性和增强维护与设计等,人工智能技术改变了传统制造流程。在本部分,我们将探讨人工智能在制造业中的应用。

The Application of AI in Manufacturing

AI is a technology that enables computers and machines to simulate human intelligence and problem-solving capabilities. On its own or combined with other technologies, AI can perform tasks that would otherwise require human intelligence or intervention.

Understanding AI in Manufacturing

AI in manufacturing is the intelligence of machines to perform humanlike tasks—responding to events internally and externally, even anticipating events—autonomously. The machines can detect a tool wearing out or something unexpected—maybe even something expected to happen—and they can react and work around the problem.

AI excels at enhancing human creativity rather than replacing it. The ideal applications enable people to focus on their unique strengths. In manufacturing, this could involve making components in a factory or designing products or parts. As AI advances, collaboration between humans and robots is becoming increasingly important. Although industrial robots are often seen as autonomous and "smart", most still require significant supervision. However, AI innovations are making these robots smarter, leading to safer and more efficient human-robot collaboration.

The Current State of AI in Manufacturing

Both large enterprises and small and medium sized enterprises (SMEs) benefit from AI adoption. Large companies have the financial strength to invest in AI, while SMEs, especially in technology-intensive industries like aerospace, are rapidly adopting new technologies to gain a competitive edge. SMEs often implement intelligent systems that provide feedback and assist in setup and operations, helping them establish a strong market presence.

Key Applications of AI in Manufacturing

AI is applied in manufacturing for various purposes, such as design and process improvement. The following are some of its applications.

➤ Factory planning and layout optimization

Facility layout must consider operator safety, process flow efficiency, and flexibility for short-run projects or frequently changing processes. Sensors track and measure space and material conflicts, and so there is a role for AI in the optimization of factory layouts.

➤ Predictive maintenance and quality inspection

AI enables predictive maintenance by equipping machines with pre-trained models that learn from data to prevent problems. AI plays a significant role in quality inspection, analyzing large data volumes from manufacturing processes. Sensors linked to AI models can flag defective parts without requiring exhaustive scanning.

➤ Generative design

AI assists in generative design, where software creates multiple design iterations based on input requirements.

➤ Flexible and reconfigurable processes

AI optimizes manufacturing processes to be more flexible and reconfigurable. AI can conduct "what-if" analysis to determine the best equipment layout and process sequencing. These AI applications help factories adapt to changing demands and potentially shift focus to different products, increasing resilience.

➢ Complementary technologies

When combined with VR and AR, AI can reduce design time and optimize assembly-line processes. Workers can use VR/AR systems for visual guidance, improving speed and precision.

➢ Intelligent components

The use of intelligent components with embedded sensors that monitor their own condition, stress, and torque is an emerging trend. This is particularly beneficial in auto manufacturing.

In the future, as humans grow AI and mature it, it will likely become important across the entire manufacturing value chain.

New Words and Phrases

simulate[ˈsɪmjuleɪt] *v.* 模仿

perform[pəˈfɔːm] *v.* 执行

anticipate[ænˈtɪsɪpeɪt] *v.* 预期,期待

replace[rɪˈpleɪs] *v.* 替换,取代

component[kəmˈpəʊnənt] *n.* 组件

supervision[ˌsuːpəˈvɪʒən] *n.* 监督

adoption[əˈdɒpʃ(ə)n] *n.* 采纳

financial[faɪˈnænʃəl] *adj.* 财务的

invest[ɪnˈvest] *v.* 投资

layout[ˈleɪaʊt] *n.* 布局

track[træk] *n.* 跟踪

flag[flæg] *v.* 标记

defective[dɪˈfektɪv] *adj.* 有缺陷的

exhaustive[ɪgˈzɔːstɪv] *adj.* 彻底的

iteration[ˌɪtəˈreɪʃ(ə)n] *n.* 迭代

reconfigurable[ˌriːkənˈfɪgjərəbl] *adj.* 可重构的

resilience[rɪˈzɪliəns] *n.* 弹性

embedded[ɪmˈbedɪd] *adj.* 嵌入的

torque[tɔːrk] *n.* 扭矩

Technical Terms

predictive maintenance 预测性维护

quality inspection 质量检查

generative design 生成式设计

"what-if" analysis "如果"分析

Augmented Reality(AR) 增强现实

Task 9 *Read the passage and match the two parts of the sentences.*

1. AI is	A. enhancing human creativity rather than replacing it.
2. AI in manufacturing is	B. a role in the optimization of factory layouts.
3. AI excels at	C. generative design, where software creates multiple design iterations based on input requirements.
4. AI enables	D. the intelligence of machines to perform humanlike tasks autonomously.
5. AI assists in	E. a technology that enables computers and machines to simulate human intelligence and problem-solving capabilities.
6. AI can conduct	F. predictive maintenance by equipping machines with pre-trained models that learn from data to prevent problems.
7. AI can reduce	G. "what-if" analysis to determine the best equipment layout and process sequencing.
8. AI plays	H. design time and optimize assembly-line processes.

Task 10 *Translate the sentences that you completed in Task 9. The first one is provided.*

1. AI is a technology that enables computers and machines to simulate human intelligence and problem-solving capabilities.

人工智能是一种使计算机和机器能够模拟人类的思考和解决问题能力的技术。

2.

3.

4.

5.

6.

7.

8.

Task 11 *Listen to a recording about digital twins. Fill in the blanks with the words provided in the box.*

• profile	• profile	• operations	• maintenance
• design	• analysis	• represent	• simulation

A digital twin is a virtual model of an object or system designed to accurately 1. ________ a physical item. It covers the entire life-cycle of the object, is updated with real-time data, and uses 2. ________, 3. ________ learning, and 4. ________ to support decision-making. Although "digital twin" is not a new term, it has become increasingly valuable with advancements in AI. It helps improve industrial operations and creates additional business value. The digital twin develops a changing 5. ________ of an asset or process in a factory, providing insights into performance throughout the plant's life-cycle, including 6. ________, 7. ________ and 8. ________.

Researching

▶ How are AI innovations transforming China's manufacturing industry? Let's study and analyze some cases together (Fig. 5. 3).

人工智能创新如何改变中国的制造业？让我们一起研究和分析一些案例。

政府出台的各项政策均提出，要以创新驱动为核心，加速人工智能等数字技术（digital technologies）的深度融合，催生新模式和新功能，助力制造业快速增长。政策不仅鼓励探索人工智能在制造业的深层次应用场景，完善智能制造的产业生态，还在如《"十四五"智能制造发展规划》（the 14th Five-Year Plan for the Development of Intelligent Manufacturing）等纲领性文件中强调，要攻克关键核心技术，研发适用于工业领域的人工智能技术，增强智能制造的融合创新能力。

Task 12 *Case Study: The Application of AI Innovations in Manufacturing in China* (*Fig. 5. 3*)

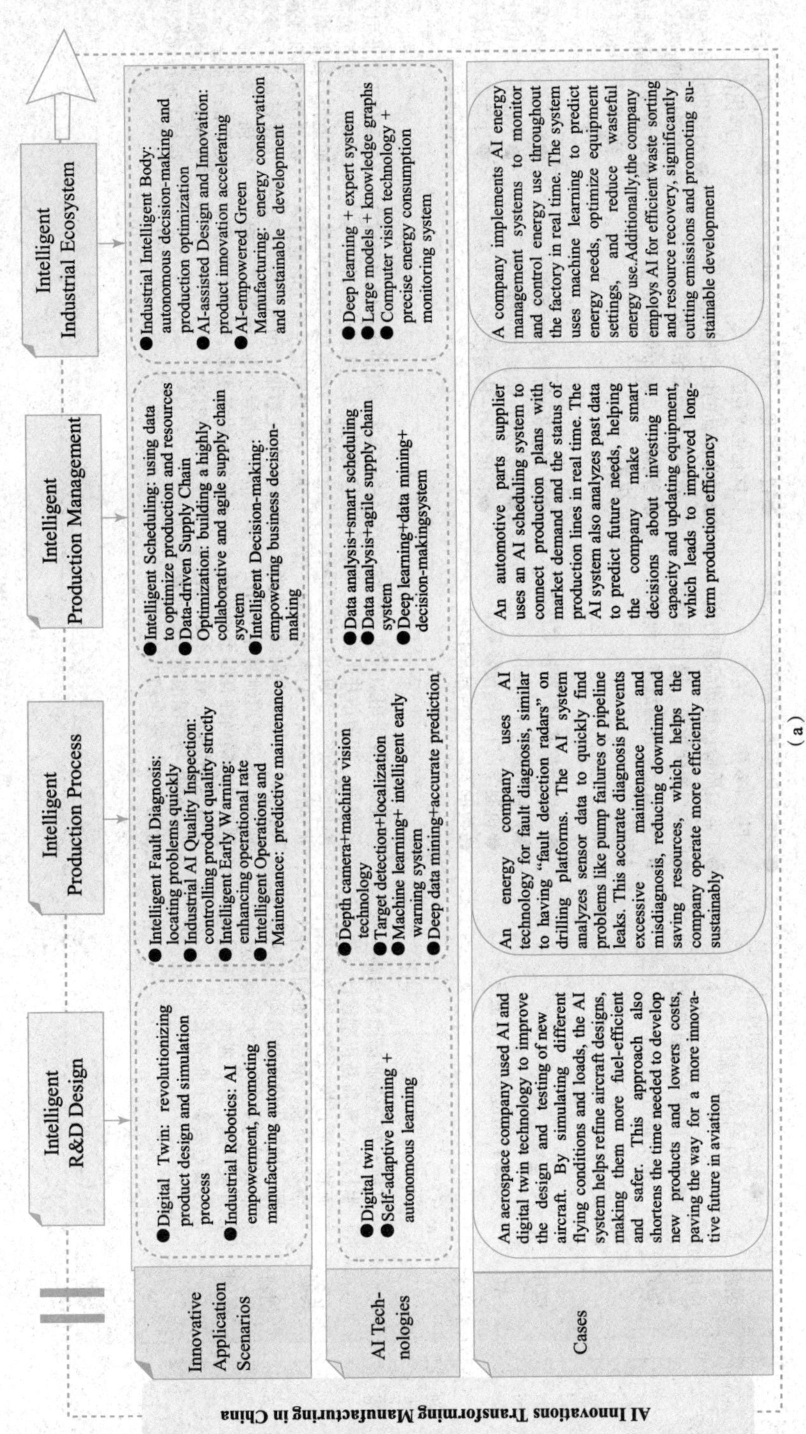

(a)

(a) English version

Fig. 5.3 AI innovations transforming manufacturing in China

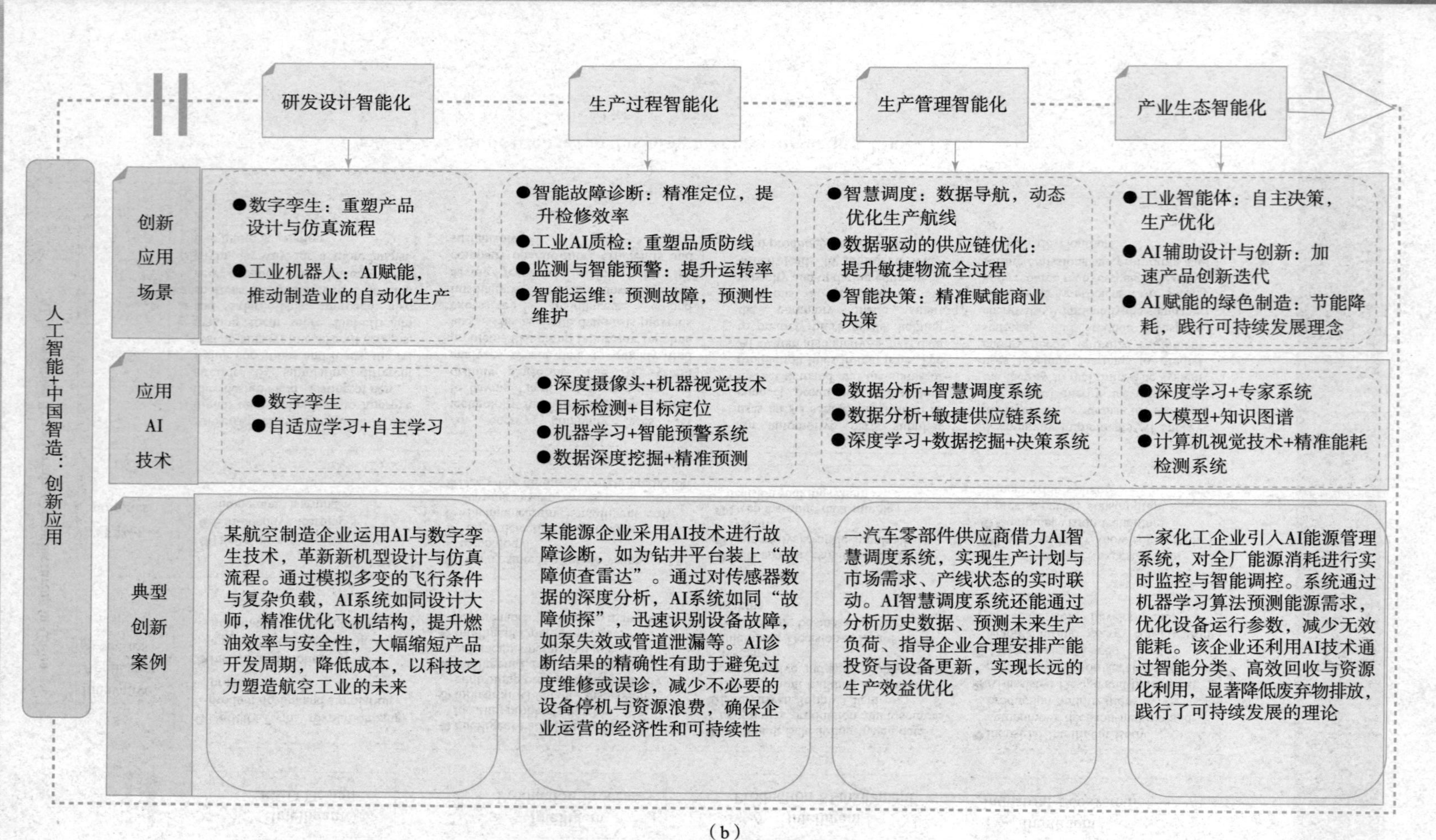

（b）

Fig. 5.3　AI innovations transforming manufacturing in China (Continued)

（b）Chinese version

Task 13 *Case Analysis: Analyze the applications of AI technologies by Chinese companies like China Railway Construction Corporation (CRCC).*

Building and Presenting

Building

实施

▶ You need to make a PPT presentation about ABC Company's innovative applications of AI in manufacturing. As a group of 4 – 6 members, you may utilize AI technology, online research, fieldwork, and the knowledge gained from this unit to better understand the topic. Then work through the following steps in Fig. 5.4 to help you complete the project.

制作一份 ABC 公司创新应用 AI 的 PPT 演示文稿,并进行汇报。请以小组(4 ~6 名成员)为单位,通过使用 AI 技术、网络调研、实地考察等方法,结合本单元所学知识,深入了解主题,并按照图 5.4 中的步骤流程完成 PPT 演示。

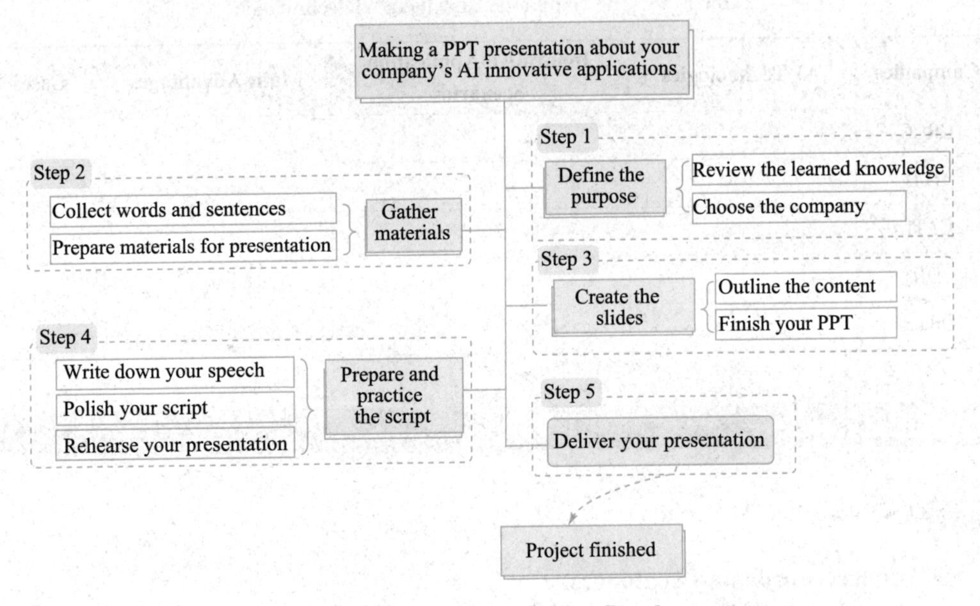

Fig. 5.4 Presentation making flowchart

Task 14

Step 1 Define the Purpose

A. Review the learned knowledge

Answer the following questions in Table 5. 1 based on what you've learned before.

Table 5. 1 List of questions under two different topics

Topics	Questions	Notes
Innovation	What is innovation?	
	How to make innovation?	
AI	What is AI?	
	What is the current state of AI in manufacturing?	
	What are the common applications?	

B. Choose the company

Search for information about the four companies in Table 5. 2, select one of them, and complete the content in the table. Or you can pick another one as you like.

Table 5. 2 The companies and their AI technologies

Companies	AI Technologies	Innovative Application Scenarios	Main Advantages	Cases
CRCC				
BYD				
CATL				
DJI				
Others				

Task 15

Step 2 Gather Materials

A. Collect words and sentences

Brainstorm words or expressions related to AI (Fig. 5. 5), and then make sentences with them.

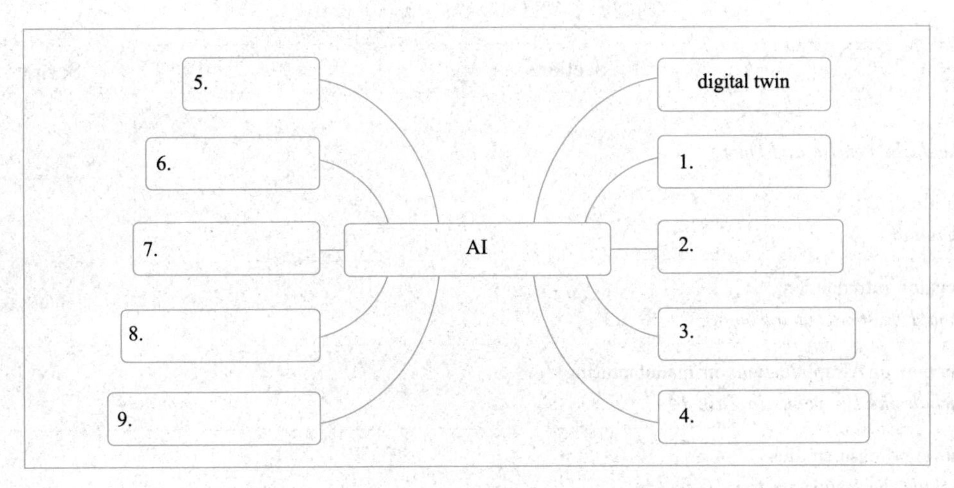

Fig. 5. 5 Words and expressions related to AI

Example: Today's focus will be on how we use AI and digital twin. Our company utilizes these technologies to simulate and optimize production processes in a virtual environment, thereby achieving greater efficiency and less downtime on the actual production line.

1.

2.

3.

4.

5.

B. Prepare materials for presentation

Search for and collect videos, audio, and images you're planning to use in your PPT. And gather data, references, and source to back up your presentation.

Task 16

Step 3 Create the Slides

A. Outline the content

Write the outline for your presentation, which should include the following parts in Table 5. 3.

Table 5.3 Sections of a presentation

Sections	Script
Title (*should be concise and clear*)	
Subtitle (*Optional*)	
Company introduction (*should be short but intriguing*)	
Overview of AI applications in manufacturing (*include the key points in Task 14*)	
Innovation case studies (*include the key points from Task 14*)	
Conclusion	

B. Finish your PPT

After finishing your outline, follow the steps in Table 5.4, and mark each one as accomplished as you go until you finish your PPT.

Table 5.4 PPT creation steps

Create a PPT	
• Select a template	( )
• Add videos or audio, images, graph, text to the template	( )
• Add animations and transition effects to the slides	( )
• Preview the effect	( )
• Make final changes and improvements	( )

Task 17

Step 4 Prepare and Practice the Script

A. Write down your speech

Write a script for your presentation, including introductions, transitions, and explanations for each slide. Apart from that, a strong start and a strong finish are essential, as shown in Table 5.5.

Table 5.5 Presentation structure

A Strong Start:
Body:
A Strong Finish:

B. Polish your script

Polish your script according to language skills you have learned in Task 2 (Table 5.6).

Table 5.6 Comparison of wording in the speech and wording after polishing

Wording in the speech	Wording after polishing

C. Rehearse your presentation

Practice makes perfect. You can practice it in front of a mirror, or record your rehearsing sessions for self-evaluation. Keep track of time and pay attention to your speech rate, tone, and body language during practice.

Presenting

展示

Task 18

Step 5 Deliver your presentation

Now it's time to give your presentation in front of the audience. There are some suggestions for you.

- Start strong: Begin with a compelling introduction that grabs your audience's attention.
- Engage your audience: Use eye contact, gestures, and voice inflection to engage your audience. Ask questions, tell stories, and use humor to keep them interested.
- End with a bang: Conclude your presentation with a strong call to action or a memorable summary of your main points.

Task 19 *After presentation, collect questions, comments, or suggestions from the audience to learn what worked well and what could be improved (Table 5.7).*

Table 5.7 Feedback from the teacher and classmates

Feedback from the teacher	
Feedback from classmates	

Task 20 *Modify the PPT and record your performance after class. The group leader sends the PPT and videos to the teacher's mailbox or uploads them to the online platform as the usual performance evaluation document and for everyone to learn from online.*

Assessing and Reflecting

Assessing

评估

▶ Assessing the work you have accomplished allows you to know not only the response of your audience to your work but also the weak parts of your work to improve.

评估你所完成的工作,不仅能了解观众对你作品的反馈,还能发现不足,以便改进。

Task 21 *The group project is evaluated jointly by teachers and students (Table 5. 8).*

Table 5. 8 Group project evaluation form

Evaluation (*Excellent* = 3, *Good* = 2, *Poor* = 1)					
Group	**Language**	**Creativity**	**Presentation**	**Collaboration**	**Cross-subject(PPT presentation)**
Group1					
Group2					
Group3					
Group4					
Group5					

Reflecting

反思

▶ Reflecting on your learning helps you understand what you've gained from this unit and how you can apply it in the future.

反思有助于你理解在本单元中所学的知识，以及将来如何加以运用。

Task 22 *The following questions in Table 5.9 may help you do the reflection, but feel free to ask yourself more questions when necessary.*

Table 5.9 Questions reflection form

Questions	Reflections
Did I increase my vocabulary to finish my task more precisely?	
Did I use any digital technologies?	
Did I actively engage in preparing the project?	
Did I develop my performing skills through PPT presentation?	
Did my contributions help others understand innovation in IM better?	
Did I identify, analyze and solve problems by fulfilling various tasks?	

Extending

► You are an engineer at a tech R&D company. Your task now is to complete virtual-real interaction in your factory.

你是某技术研发公司的工程师，现在需要在实际生产线中复现虚实联动。

Task 1 *Get acquainted with industrial digital twin.*

The industrial digital twin is the application of digital twin technology in the industrial sector, as shown in Fig. 5.6. It not only enhances production efficiency and reduces costs, but also optimizes product design and production processes through simulation and prediction, achieving more flexible and personalized production.

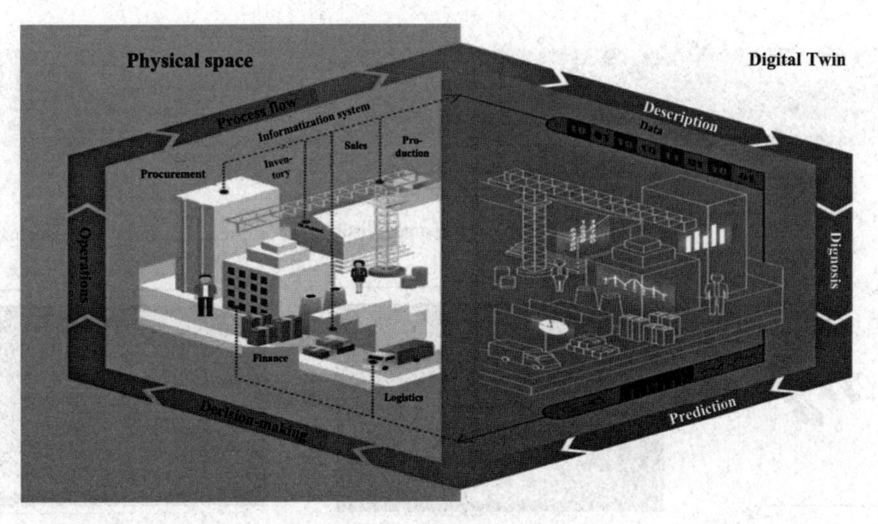

Fig. 5. 6 Industrial digital twin

Task 2 *Familiarize yourself with the structure of the digital twin system (Table 5. 10).*

Table 5. 10 Digital twin system

Digital Twin System		
Architecture Layer	Key Technologies	Main Functionalities
Physical equipment	Production management systems	Support for the physical layer: real-time data capture and analysis support for the physical layer in digital twin system
Data transmission	Data transmission protocols, databases	Data collection: process environment, physical characteristics of the equipment, operation patterns
Industrial Internet data platform	Data warehouse, visualization management	Visual display: 3D scene simulation
Digital twin body	Model displacement animation, model vertex animation, model skeletal animation, AI target filtering, control callback, calculating physical entity, model adding/deleting, scene highlighting/blurring, special effects display, graphical AI computation	Process for digital twin implementation: • Construct animation for layers of physical model nodes • Control animation of physical models • Control displacement of physical models • Control algorithms of physical models • Implement digital twin technology

Task 3 *Complete the operation process in Fig. 5. 7 based on Task 1 and Task 2.*

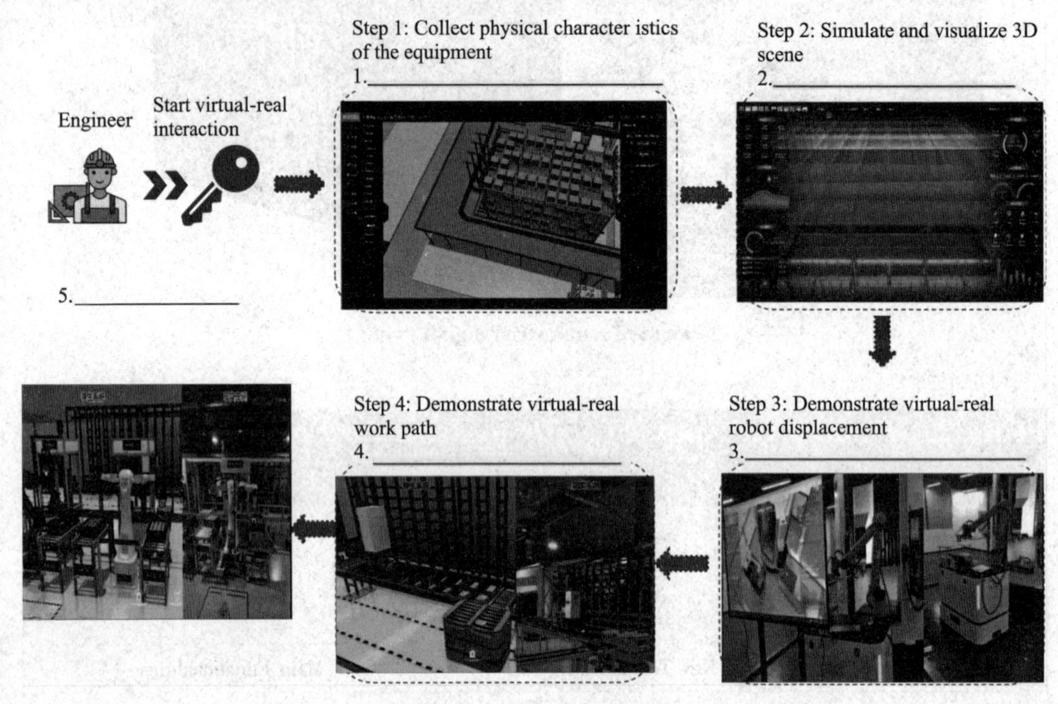

Fig. 5. 7 Operation flowchart

Unit 6 Transformation towards Intelligent Manufacturing 向智能制造转型

Unit Introduction

Driving Question

驱动问题

The path towards high-end, intelligent, and green manufacturing is becoming increasingly clear. Therefore, it is crucial to accelerate digital and green transformations. How can a manufacturing company successfully transform to adapt to new technological changes, enhance overall competitiveness, and achieve sustainable development?

制造业正朝着高端化、智能化和绿色化的方向发展，因此中国必须加快数字化和绿色转型的步伐。那么制造公司如何成功转型，以适应新技术变革、提升整体竞争力、实现可持续发展呢？

Project Overview

项目概述

In this unit, you will begin by exploring the basic concepts of digital transformation (DX). After that, you will delve into the concept of green manufacturing and how it aligns with digital transformation goals. Next, you will learn about new quality productivity forces. Then, you will study practical application cases of Chinese companies that have successfully implemented digital transformation strategies. Next, working in groups, you will deliver a public speech to explain ABC Company's digital transformation journey. After the speech, you will discuss the benefits and challenges of transformation.

在本单元，首先学习数字化转型的基本概念，然后深入学习绿色制造的概念及其与数字化转型的关系，学习新质生产力的概念。接下来，研究成功地实施了数字化转型战略的中国企业的实际应用案例。随后，小组合作发表有关 ABC 公司数字化转型经验的演讲。最后，讨论转型的好处和挑战。

Project Extension

项目拓展

After completing the main project, you'll engage in a hands-on extension activity to deepen your understanding of the transformation of Chinese manufacturing industry. This will involve learning how to analyze the trends of Chinese manufacturing going abroad.

在完成主要项目后,将进行实操拓展,分析中国制造业出海的全球趋势,以进一步加深对中国制造业转型的理解。

Learning Objectives
学习目标

This unit is intended to help you:

1. have a general idea of digital transformation;
2. explore methods for green manufacturing and sustainable development;
3. analyze successful cases of digital transformation in English;
4. deliver an English speech on digital transformation;
5. analyze the trends of Chinese manufacturing going abroad;
6. cultivate the awareness of sustainable development.

本单元旨在帮助你:

1. 了解数字化转型;
2. 探讨实现绿色制造和可持续发展的路径;
3. 用英语简单分析企业实施数字化转型的成功案例;
4. 发表关于数字化转型的英文公共演讲;
5. 分析中国制造业出海的趋势;
6. 培养可持续发展意识。

Getting Ready

▶ What is a public speech, and how to use language vividly in a speech? Please complete the tasks below to find the answer. Fig. 6. 1 shows a man giving a public speech.

什么是公众演讲?如何在演讲中运用生动的语言?请完成以下任务,找到答案。图 6. 1 显示的是一位男士正在进行公众演讲。

Fig. 6. 1 Public speech

Task 1 *Listen to a recording about what a speech is and answer the following questions.*

1. What is the role of speech in communication?

Speech is a critical mode of communication, whether verbal or non-verbal, and it is used to ________________. It is crucial for various professions as it facilitates ________________.

2. Is the ability to give a good keynote speech something one is born with?

No, the ability to give a good keynote speech is not an innate talent. It is a skill that can be developed over time ____.

3. It is common to feel nervous before giving a speech, and how can one overcome this?

The key to overcoming this lies in one's ________________, ____________________ fosters the necessary confidence.

4. How can body language contribute to the effectiveness of a speech?

Body language, such as ______________, ________________, and ________________, ________________ conveys authority and helps command the audience's attention, thus contributing to the effectiveness of a speech.

5. What is the significance of vocal tone when delivering a speech?

The vocal tone is significant because speaking with conviction ______________________ ____. It helps to project ________________________, which is important in motivational or leadership speeches.

Task 2 *Learn some figures of speech.*

演讲中适当使用一些修辞手法，可以使演讲更加生动、有趣，更具有说服力。

2.1 明喻(Simile)/暗喻(Metaphor)

特点：将两个在本质上不同，而却有着相似之处的事物做一个显性比较，帮助听众理解复杂概念。在结构上明喻包含 like 或 as，暗喻不包含。

示例：Our new software is like a Swiss Army knife for businesses, versatile and indispensable.

2.2 排比(Parallelism)

特点：通过将一对或一系列词语、词组或句子用相似的结构排列在一起，强调产品优势，增强语言的节奏感与说服力。

示例：Our product is fast, efficient, and user-friendly, offering unparalleled performance in its class.

2.3 对照(Antithesis)

特点：有意识地把意义相对的词语放在对称的结构中，增添语言的吸引力。

示例：While other solutions may be bulky and slow, our device is sleek, speedy, and ahead of its time.

2.4 反问(Rhetorical question)

特点：通过提问激发听众思考，可增强听众的参与感或强调演讲者的观点。

示例：Who hasn't struggled with outdated technology? Well, the wait for a better solution is over.

2.5 重复(Repetition)

特点：在连续的句子、词组的开头或结尾重复同一个或同一组词语，来强调主题。

示例:	Innovative, innovative, innovative-our approach to design is what sets us apart.

Task 3 *Identify the rhetorical figures used in the following sentences.*

1. The assembly line is a river flowing through the heart of our operations. (   )

2. We test, we evaluate, we improve. (   )

3. Our research team is a beacon of innovation in the industry. (   )

4. Is there any reason not to prioritize green and environmentally friendly practices in our manufacturing processes? (   )

5. Can we achieve higher efficiency without improving our processes? (   )

6. In the past, production efficiency was often achieved at the expense of the environment, but now we strive for a balance where efficiency and sustainability go hand in hand. (   )

7. We grow here, we grow there, we grow everywhere. (   )

8. In the bustling marketplace of today, where innovation and competition are the driving forces, we find a stark contrast to the simple barter systems of the past, where trade was a matter of necessity rather than a platform for growth and expansion. (   )

9. We satisfy, we delight, we exceed expectations. (   )

10. Can there be a future where artificial intelligence doesn't play a pivotal role in shaping our lives and transforming our world? (   )

11. AI, it's the future; AI, it's the present; AI, it's the key to unlocking endless possibilities. (   )

12. Intelligent manufacturing is like a clever chef, carefully blending technology and processes to cook up efficient and high-quality production. (   )

Starting up

▶ Digital transformation is crucial for manufacturers to stay ahead in a competitive market. But what does digital transformation entail for the manufacturing industry? In this section, we will explore its significance, main types and the impact it has on modern manufacturing practices.

数字化转型对企业在竞争激烈的市场中保持领先至关重要。那么,数字化转型究竟意味着什么?在这一部分,我们将探讨其重要性、主要类型及其对制造业的影响。

Digital Transformation in Manufacturing

Digital transformation is a concept that is often mixed up with two other ideas: digitization and digitalization.

Digitization is the simplest form and involves changing information from analog or manual form to digital form. Digitalization means using digital technology to improve how you do things. It's about doing your regular tasks in new, more efficient ways that lead to better results. Digitalization lays the groundwork for digital transformation, which allows businesses to rethink how they use technology, people, and processes to advance their operations.

Digital Transformation in Manufacturing

Digital transformation in manufacturing focuses on using new technology to make operations more efficient and effective. This typically includes high-tech tools like digital twin, IoT sensors, and AI to monitor and automate processes.

By adopting these technologies, manufacturers can gain faster insights across their entire facility, improving everything from daily manufacturing tasks to supply chain management. In this way, companies can enhance efficiency, ensure safety, and support growth.

Five Examples of Manufacturing Digital Transformation

In manufacturing, there are five types of transformation and often they are done in some sort of combination.

➢ Process transformation

In this type of transformation, businesses examine their processes—whether that's in the design process, the factory floor, or field service—and implement digital technology to support process change.

Example business initiatives are implementing digital performance management within the factory, optimizing design for manufacturability and serviceability, and improving collaboration across the product lifecycle.

➢ Product and service transformation

To remain competitive in a changing market, companies must offer innovative products and high-quality service. Digital transformation helps companies quickly develop new products, features, and customizations that customers want and find ways to enhance service delivery.

Examples of digital transformation initiatives include creating a digital thread that provides real-world product usage data to design teams for improvement, enhancing product quality, and enabling predictive maintenance and remote service.

➢ Growth opportunities

Manufacturing companies use digital transformation to find and pursue growth opportunities in various ways. Common strategies include improving factory efficiency to increase output, developing new service models or features, and better managing the complexity of products and portfolios.

➢ Customer experience

In a competitive market, it is essential to meet customer needs and desires. Using technology to enhance customer experiences and engagement can set a company apart.

Customers expect more from products, particularly in terms of technology. Offering software-driven innovations is a key way to differentiate products and provide a unique customer experience.

➢ Cultural transformation

This is essential for any digital transformation. Employees are expected to adopt new technologies and processes, which can initially seem challenging. Building a culture that encourages collaboration and clearly communicates the business goals of digital transformation is crucial for success.

Every business in today's global economy can benefit from digital transformation in some way. We will flourish if we embrace a process of ongoing renewal-always striving to understand what customers need, the direction our market is heading, and how technology can give us a competitive advantage.

New Words and Phrases

digitization[ˌdɪdʒɪtə'zeɪʃn]n. 数字化

digitalization[ˌdɪdʒɪtələ'zeɪʃn]n. 数字化进程

analog['ænəlɒg]n. 模拟信号

groundwork['graʊndwɜːk]n. 基础

rethink['riːθɪŋk]v. 重新考虑

deliver[dɪ'lɪvə]v. 传递

automate['ɔːtəmeɪt]v. 自动化

combination[ˌkɒmbɪ'neɪʃ(ə)n]n. 结合

remote[rɪˈməʊt] *adj.* 远程的

output[ˈaʊtpʊt] *n.* 产量

portfolio[pɔːtˈfəʊliəʊ] *n.* 组合

differentiate[ˌdɪfəˈrenʃieɪt] *v.* 区分

flourish[ˈflɜːrɪʃ] *v.* 繁荣

strive[straɪv] *v.* 努力,奋斗

mix up with　混淆

lay the groundwork for　为……奠定基础

regardless of　不管

set... apart　使突出

strive to　努力

Technical Terms

digital twin　数字孪生

IoT sensors　物联网传感器

digital performance management　数字绩效管理

digital thread　数字线程

remote service　远程服务

Task 4　*According to the passage, are the following statements true (T) or false (F)?*

(　　)1. Digital transformation is the same as digitization.

(　　)2. Digitalization is about changing information from analog to digital form.

(　　)3. Digital twin, IoT sensors, and AI are examples of high-tech tools used in manufacturing digital transformation.

(　　)4. Process transformation focuses on improving customer experience and engagement.

(　　)5. Digital transformation does not involve improving product quality.

(　　)6. Predictive maintenance and remote service are examples of digital transformation initiatives that enhance service delivery.

(　　)7. Cultural transformation is not necessary for a company to achieve the success of digital transformation.

(　　) 8. To remain competitive in a changing market, companies must offer cheap products.

Task 5 *Find the answers to the following questions in the passage.*

1. What is the main difference between digitization and digitalization?

____

2. What is digital transformation in manufacturing?

____

3. How can manufacturing companies find and pursue growth opportunities?

____

4. Give an example of process transformation in manufacturing digital transformation.

____

5. How does digital transformation help in improving customer experience?

____

Task 6 *Match the English expressions in Column A with their Chinese meanings in Column B.*

Column A	Column B
1. digital twin	A. 数字绩效管理
2. IoT sensor	B. 远程服务
3. digital performance management	C. 数字线程
4. digital thread	D. 数字孪生
5. remote service	E. 数字化转型
6. customer experience	F. 产品组合
7. product portfolio	G. 客户体验
8. digital transformation	H. 物联网传感器

Task 7 *Make six sentences with words from the following boxes.*

transformation	digitization	digitalization	output
groundwork	initiative	portfolio	combination

Example: The shift towards sustainable manufacturing practices has sparked a green **transformation** in the industry, reducing waste and environmental impact.

1. ____.
2. ____.
3. ____.
4. ____.
5. ____.
6. ____.

Task 8 *Translate the following sentences into English using the words and expressions given.*

1. 开发客户关系管理系统为改善客户服务奠定基础。(lay the groundwork for)

____.

2. 培养阅读习惯至关重要。(It is essential)

____.

3. 无论员工在哪里工作,公司都有义务保护他们。(regardless of)

____.

4. 这些品牌深谙如何实现与竞争对手的差异化发展。(differentiate)

____.

5. 天行健,君子以自强不息。(strive)

____.

Investigating

► Under the vision of achieving "carbon neutrality and carbon peaking", the dual transformation of digitalization and greening is an inevitable choice to adapt to new changes in the technological environment and help form new quality productive forces. In this section, we will explore what green manufacturing is.

在“双碳”愿景下,数字化、绿色化双转型是适应技术环境新变化,助力形成新质生产力的必然选择。在这一部分,我们将探讨什么是绿色制造。

Green Manufacturing

Manufacturing is the backbone of China's national economy. However, manufacturing facilities are also responsible for a large share of the country's carbon emissions and energy usage.

To minimize the impact manufacturing has on the environment, many plants are adopting more sustainable and eco-friendly practices. These efforts are not only beneficial for the planet but also help companies enhance their profitability.

Understanding Green Manufacturing

Green manufacturing, also known as sustainable manufacturing or green production, focuses on minimizing environmental impact and maximizing resource efficiency. This approach involves renewing production processes to use fewer natural resources, reduce pollution and waste, and moderate emissions.

By adopting eco-friendly technologies and practices, green manufacturers lower their energy consumption and overall ecological footprint. These efforts not only benefit the environment but also improve operational efficiency and cost-effectiveness, making sustainability more accessible for businesses.

Five Ways to Implement Green Manufacturing

➢ Adopt lean manufacturing principles

Adopt lean manufacturing principles to improve operations and reduce waste. By simplifying processes, you can use resources more efficiently. Implementing practices like just-in-time inventory management and continuous improvement helps minimize waste and enhance sustainability.

➢ Invest in energy-efficient technologies

Invest in energy-efficient technologies by upgrading your equipment and machinery. Use modern options like high-efficiency motors, solar panels, smart grid technologies, and IoT sensors for energy monitoring. Implementing AI-driven energy management systems and smart building technologies that automatically adjust lighting and heating, ventilation, and air conditioning systems. Reducing energy consumption lowers costs and decreases your carbon footprint, contributing to a greener future.

➢ Optimize supply chain management

Work with your suppliers to get sustainable materials and components. Choose suppliers who follow eco-friendly practices, use recycled materials, and are committed to ethical sourcing. By making your supply chain more sustainable, you can ensure your products are environmentally friendly from start to finish.

➢ Implement waste reduction strategies

Implement waste reduction strategies to create less waste and encourage recycling and reuse. Use efficient production methods to find ways to reduce waste and improve processes. Consider innovative waste management solutions like closed-loop systems and material recovery programs to reduce environmental impact and increase resource efficiency.

➢ Educate and empower employees

Educate and empower your employees to create a culture of sustainability. Offer training and awareness programs on eco-friendly practices and the importance of sustainability. Encourage employees to participate in initiatives like energy conservation and waste reduction. By building a team committed to sustainability, you can drive positive change and contribute to a greener future.

Taking steps to become more sustainable can have a positive impact on any business. It helps to improve the brand image, creates a cleaner, safer work environment, and helps the company become more profitable. Protecting the planet has never been more convenient or more productive.

New Words and Phrases

backbone[ˈbækbəʊn] *n.* 支柱

responsible[rɪˈspɒnsəb(ə)l] *adj.* 有责任的

emission[iˈmɪʃ(ə)n] *n.* 排放
minimize[ˈmɪnɪmaɪz] *v.* 最小化
beneficial[ˌbɛnɪˈfɪʃəl] *adj.* 有益的
profitability[ˌprɒfɪtəˈbɪlɪti] *n.* 赢利
maximize[ˈmæksɪmaɪz] *v.* 最大化
renew[rɪˈnjuː] *v.* 更新
consumption[kənˈsʌmpʃ(ə)n] *n.* 消耗
ecological[ˌiːkəˈlɒdʒɪk(ə)l] *adj.* 生态的
footprint[ˈfʊtˌprɪnt] *n.* 碳足迹
lower[ˈləʊə(r)] *v.* 降低,减少
lean[liːn] *adj.* 精简的
simplify[ˈsɪmplɪfaɪ] *v.* 简化
ventilation[ˌvɛntɪˈleɪʃ(ə)n] *n.* 通风
conservation[ˌkɒnsəˈveɪʃ(ə)n] *n.* 保护

be responsible for 对……负责
have impact on 对……有影响

Technical Terms

green manufacturing 绿色制造
sustainable manufacturing 可持续制造
ecological footprint 生态足迹
smart grid technologies 智能电网技术
closed-loop systems 闭环系统

Task 9 *Read the passage and match the two parts of the sentences.*

1. Manufacturing is	A. the backbone of China's national economy.
2. Green manufacturing focuses on	B. suppliers who follow eco-friendly practices, use recycled materials, and are committed to ethical sourcing.
3. Green manufacturers lower	C. minimizing environmental impact and maximizing resource efficiency.

Continued

4. Manufacturers can adopt	D. waste reduction strategies to create less waste and encourage recycling and reuse.
5. Manufacturers can invest in	E. employees training and awareness programs on eco-friendly practices and the importance of sustainability.
6. Manufacturers can choose	F. lean manufacturing principles to improve operations and reduce waste.
7. Manufacturers can implement	G. their energy consumption and overall ecological footprint by adopting eco-friendly technologies.
8. Companies can offer	H. energy-efficient technologies by upgrading your equipment and machinery.

Task 10 *Translate the sentences that you completed in Task 9. The first one is provided.*

1. Manufacturing is the backbone of China's national economy.
制造业是国民经济的命脉。

2.

3.

4.

5.

6.

7.

8.

Task 11 *Listen to a recording about new quality productive forces. Fill in the blanks with the words provided in the box.*

• iteration	• inseparable	• highlighted	• innovation
• intrinsic	• efficiency	• quality	• modernization

The new quality productive forces 1. ________ in the 2024 government work report have attracted much attention. Essentially, new quality productive forces refer to advanced productivity characterized by 2. ________, with a core focus on the excellence of 3. ________. It features high technology, high 4. ________ and high quality, and aligns with the country's new development philosophy. Developing these forces is an 5. ________ requirement for high-quality development and a key to promoting the 6. ________ and upgrading of productivity, as well as realizing 7. ________. The development of projects such as new energy vehicles, 5G base stations, intelligent robots, and mobile payments is 8. ________ from the contributions of new quality productive forces.

Researching

▶ How have companies in China successfully implemented digital transformation? Let's explore and analyze some cases together.

中国企业如何成功实施数字化转型？让我们一起研究和分析一些案例。

近年来，中国制造业数字化转型逐渐深化，数字技术赋能制造业快速推进。制造业在数字化转型过程中，生产力构成要素的数字化变革发生新的跃进，进一步形成新质生产力，也为绿色工业体系的发展提供了技术支撑。

Task 12 *Cases Study: Digital Transformation in Manufacturing in China (Fig. 6.2).*

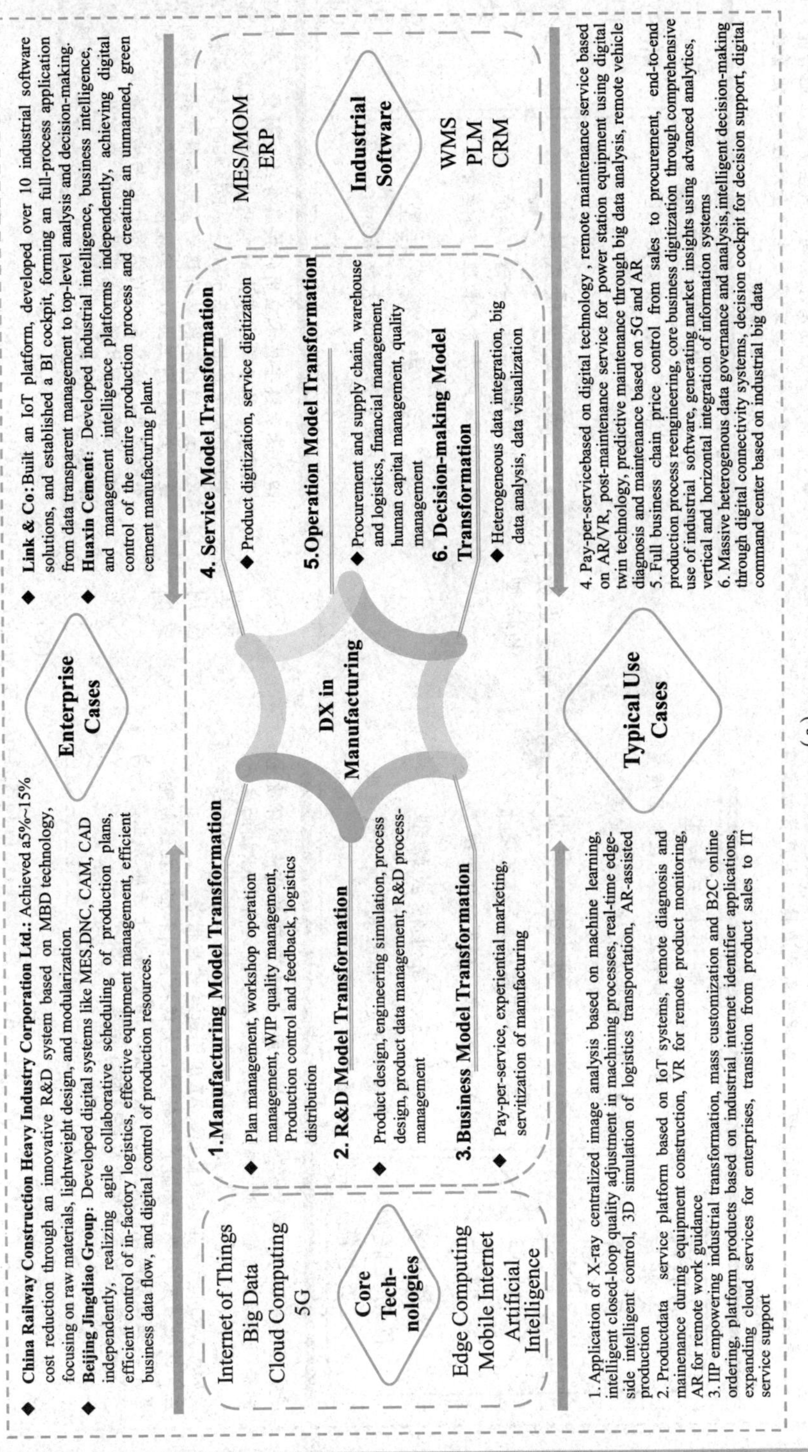

(a)

Fig. 6.2 Digital transformation in manufacturing in China

(a) English version

(b)

Fig. 6.2 Digital transformation in manufacturing in China (Continued)

(b) Chinese version

Task 13 ***Case Analysis: Analyze the main types of digital transformation employed by successful Chinese companies like Beijing Jingdiao Group.***

Building and Presenting

Building

实施

▶ You need to deliver a public speech at the conference to explain ABC Company's digital transformation. As a group of 4 – 6 members, you may utilize AI technology, online research, fieldwork, and the knowledge gained from this unit to better understand the topic. Then work through the following steps in Fig. 6. 3 to help you complete the project.

发表公众演讲,讲解 ABC 公司的数字化转型。请以小组(4 ~6 名成员)为单位,通过使用 AI 技术、网络调研、实地考察等方法,结合本单元所学知识,深入了解主题,并按照图 6. 3 中的步骤流程完成公众演讲。

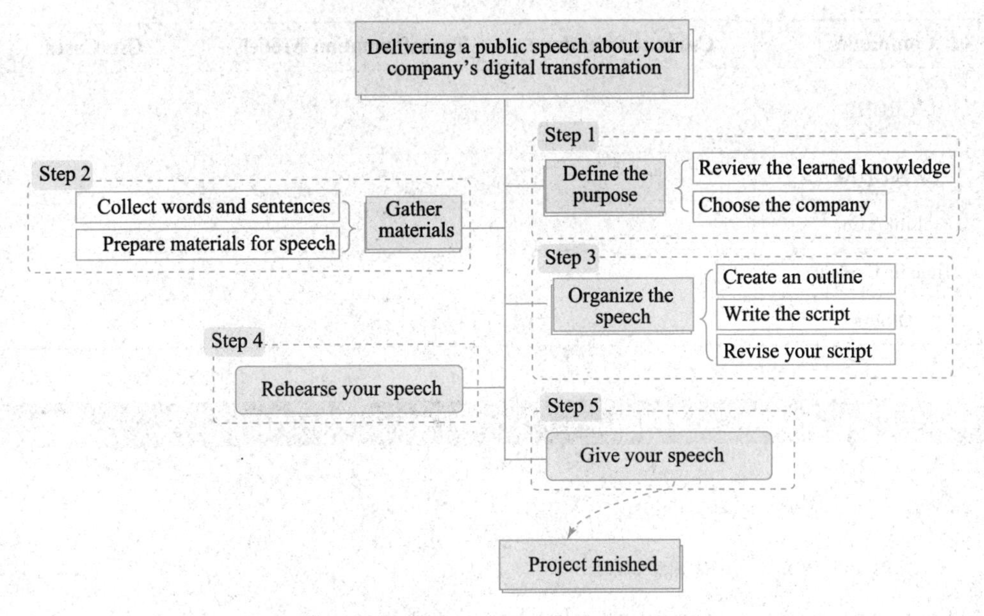

Fig. 6. 3 Speech delivery process flowchart

Task 14

Step 1 Define the Purpose

A. Review the learned knowledge

Answer the following questions in Table 6.1 based on what you've learned before.

Table 6.1 List of questions under two different topics

Topics	Questions	Notes
Digital transformation	What is digital transformation?	
	What are the differences between digitization and digitalization?	
	List the types of digital transformation in manufacturing.	
Green manufacturing	What is green manufacturing?	
	How to implement green manufacturing?	

B. Choose the company

Search for information about the four companies in Table 6.2, select one of them, and complete the content in the table. Or you can pick another one as you like.

Table 6.2 The Companies and their situations

Companies	Current Situation	Transformation Models	Use Cases
CRCHI			
WEICHAI			
Link&Co			
Huaxin Cement			
Others			

Task 15

Step 2 Gather Materials

A. Collect words and sentences

Brainstorm words or expressions related to digital transformation (Fig. 6.4), and then make sentences with them.

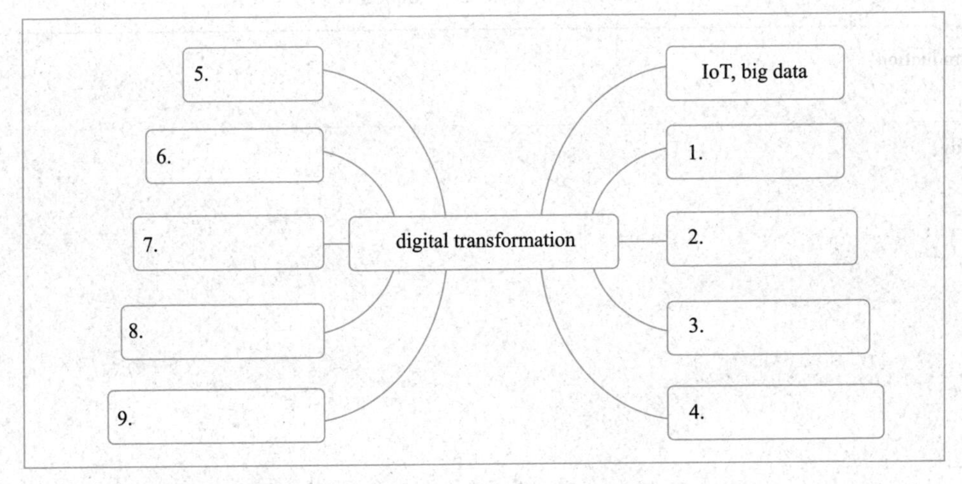

Fig. 6.4 Words and expressions related to digital transformation

Example: Our company has forged a cutting-edge IoT platform, crafted over ten industrial software solutions, and built a powerful engine.

1.

2.

3.

4.

5.

6.

B. Prepare materials for speech

Do research on your company's digital transformation cases. And then search for and collect data, references, and source you're planning to use in your speech.

Task 16

Step 3 Organize the Speech

A. Create an outline

Outline your speech's structure. Think about the main ideas for each section and finish Table 6.3.

Table 6.3 Speech's structure

Introduction:
Body:
Conclusion:

B. Write the script

Compose the script for your speech according to the outline (Table 6.4).

Table 6.4 Sections of a speech and writing requirements

Sections	Script
Introduction (*include the self-introduction and company introduction*)	
Overview of company's digital transformation (*include the key points from Task 14*)	
Digital transformation case studies (*include the key points from Task 14*)	
Conclusion (*include future look*)	

C. Revise your script

Polish your script according to language skills you have learned in Task 2 (Table 6.5).

Table 6.5 Comparison of wording in the script and Wording after polishing

Wording in the script	Wording after polishing

Task 17

Step 4 Rehearse Your Speech

Practice makes perfect. You can practice it in front of a mirror, or record your rehearsing sessions for self-evaluation. Keep track of time and pay attention to your body language, pace, volume, and tone of voice.

Presenting

展示

Task 18

Step 5 Give Your Speech

Now it's time to give your speech in front of the audience. There are some suggestions for you.

- Be well prepared and confident.
- Engaging your audience.
- Preparing for Q&A.

Task 19 *After the speech, collect questions, comments, or suggestions from the audience to learn what worked well and what could be improved (Table 6.6).*

Table 6.6 Feedback from the teacher and classmates

Feedback from the teacher	

Continued

Feedback from classmates	

Task 20 *Modify the script and record your performance after class. The group leader sends the videos to the teacher's mailbox or uploads them to the online platform as the usual performance evaluation document and for everyone to learn from online.*

Assessing and Reflecting

Assessing

评估

► Assessing the work you have accomplished allows you to know not only the response of your audience to your work but also the weak parts of your work to improve.

评估你所完成的工作,不仅能了解观众对你作品的反馈,还能发现不足,以便改进。

Task 21 *The group project is evaluated jointly by teachers and students (Table 6.7).*

Table 6.7 Group project evaluation form

Evaluation (*Excellent* = 3, *Good* = 2, *Poor* = 1)					
Group	**Language**	**Creativity**	**Presentation**	**Collaboration**	**Cross-subject(speech)**
Group1					
Group2					
Group3					
Group4					
Group5					

Reflecting

反思

▶ Reflecting on your learning helps you understand what you've gained from this unit and how you can apply it in the future.

反思有助于你理解在本单元中所学的知识，以及将来如何加以运用。

Task 21 *The following questions in Table 6.8 may help you do the reflection, but feel free to ask yourself more questions when necessary.*

Table 6.8 Questions reflection form

Questions	Reflections
Did I increase my vocabulary to finish my task more precisely?	
Did I use any digital technologies?	
Did I actively engage in preparing the project?	
Did I develop my performing skill through giving a speech?	
Did my contributions help others understand transformation towards IM better?	
Did I identify, analyze and solve problems by fulfilling various tasks?	

Extending

▶ You are the head of international business at a Chinese high-tech manufacturing company. Your task is to analyze the trends of Chinese manufacturing going global, identifying market opportunities and challenges. This analysis will support the company's future international market strategy.

你是一家中国高科技制造企业的国际业务负责人。你的任务是分析中国制造业出海的全球趋势，识别市场机会和挑战，为开拓公司未来的国际市场提供决策支持。

Task 1 *Get acquainted with Chinese enterprises "going global".*

As China's reform and opening up continues, the country is seeing a growing capability in supply of products and services. The volume and types of China-manufactured commodities and Chinese services continue to expand, and the gradually optimized structure of the sector is also leading to a broader market, making the path of "going global" wider and wider.

The commodity structure in the Chinese manufacturing industry is also being continuously optimized, as high-tech products have replaced labor-intensive ones as the major force of exports. "Made in China" products are destroying their old image of being low-quality and simply assembled, and more and more fashionable Chinese products are being exported to foreign markets.

Task 2 *Familiarize yourself with "KPMG's 2023 White Paper on Chinese Manufacturing Enterprises Going Global".*

2.1 China is a global manufacturing leader

According to the World Bank's data in 2022, China's manufacturing value-added has reached nearly $5 trillion, representing 30.7% of the global total. Fig. 6.5 indicates a consistent increase in China's contribution to global manufacturing, both in terms of absolute value and global share, peaking in 2021. The bar graph represents China's manufacturing value-added in trillions of USD and the line graph shows China's share of global manufacturing value-added as a percentage over the same period. The value-added in USD increases steadily from 2016 to 2021, reaching its peak in 2021, slightly decreasing in 2022 and 2023 but still remaining above $4 trillion. The percentage share of global manufacturing value-added also shows a steady increase from approximately 27% in 2016 to over 30% by 2021, maintaining around 30% in 2022 and 2023.

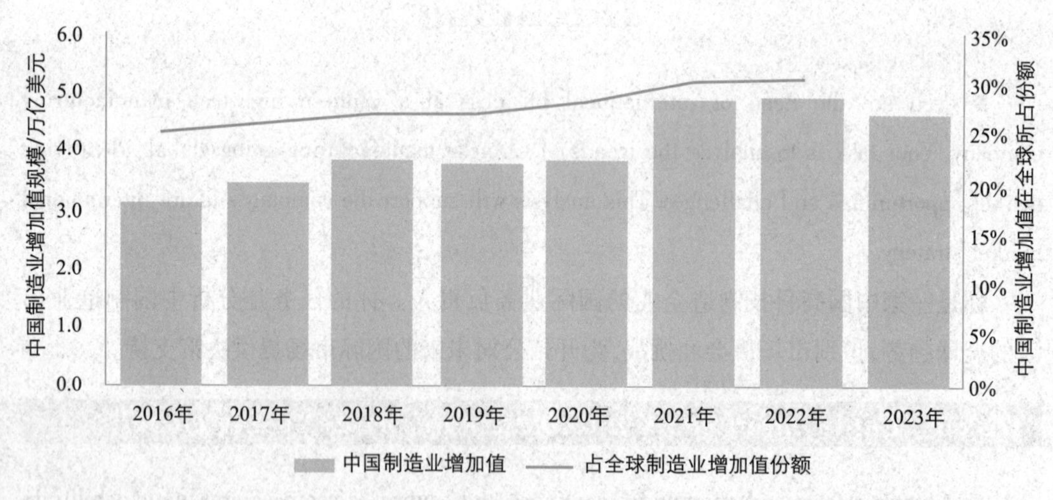

Fig. 6.5 China's manufacturing scale and global share

2.2 Structure of key products exported in manufacturing

China's manufacturing exports continue to shift from labor-intensive to technology-intensive products. The share of mechanical and electrical products and high-tech products in China's exports has been steadily rising. In particular, the export share of mechanical and electrical products increased from 58.6% in 2019 to 74.4% in 2023, while the share of high-tech products fluctuated around 30%. In contrast, the export of labor-intensive products has been decreasing slightly, with labor-intensive products accounting for 24.2% of total exports in 2023, as shown in Fig. 6.6.

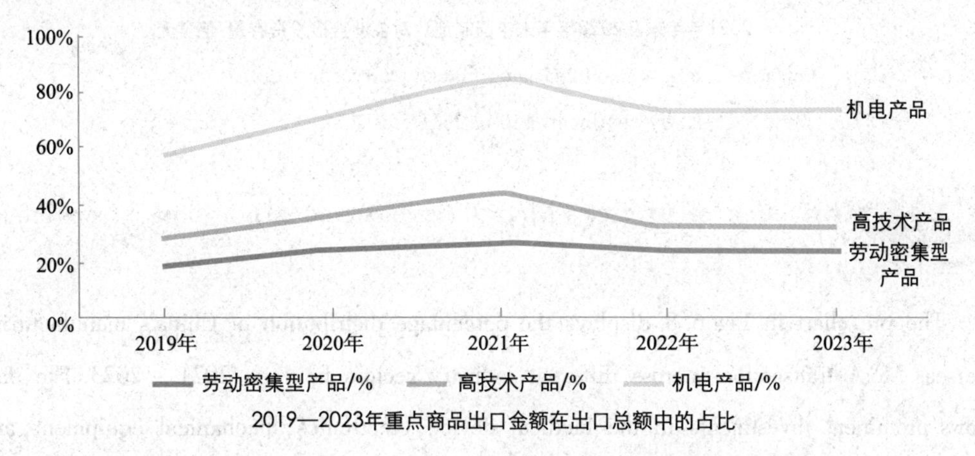

Fig. 6.6 The proportion of export amounts for labor-intensive products, high-tech products, and mechanical and electrical products as part of total exports from 2019 to 2023

2.3 China's outbound investments in manufacturing

The bar chart in Fig. 6.7 presents data on investment stock by region for the end of 2021 and 2022, emphasizing Asia's dominant position, with increasing investments in Europe and significant investments in Latin America and North America. In terms of regional investments, China's manufacturing industry continues to focus on Asia, where by the end of 2022, the stock of investments had reached \$136.13 billion, ranking first. Meanwhile, investment in Europe has shown an upward trend, with a 16.7% increase in 2022 compared to the previous year. China's manufacturing stock investment in Latin America also remained high at \$47.74 billion, ranking third, followed by North America with \$28.66 billion by the end of 2022.

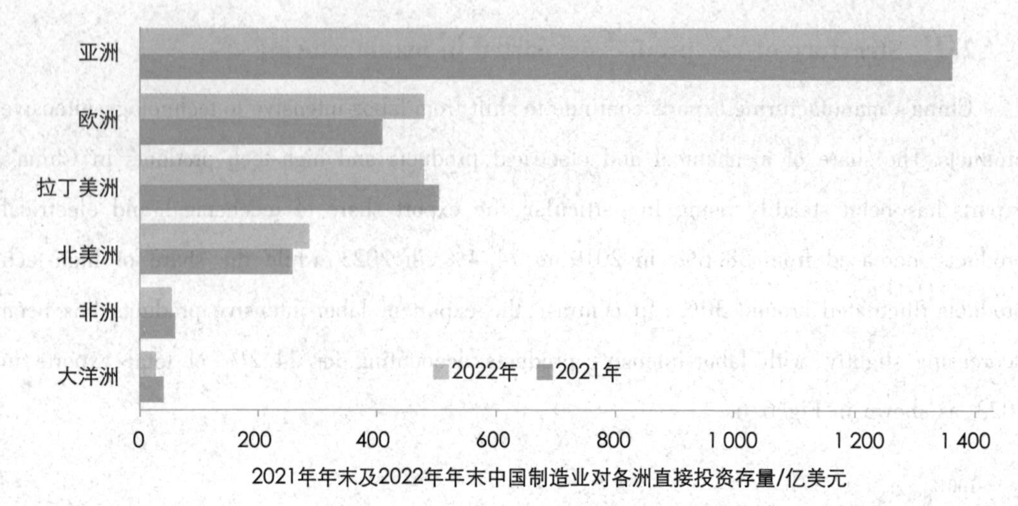

Fig. 6. 7 China's manufacturing industry direct investment stock by continent(end of 2021 and 2022)

2. 4 Distribution of overseas M&A (mergers and acquisitions) transactions in manufacturing

The pie chart in Fig. 6. 8 displays the percentage distribution of China's manufacturing overseas M&A transactions across different industry sectors between 2021 – 2023. The data shows prominent investments in the medical device, electronics, mechanical equipment, and automotive industries. In specific manufacturing sub-sectors, Chinese companies mainly invested

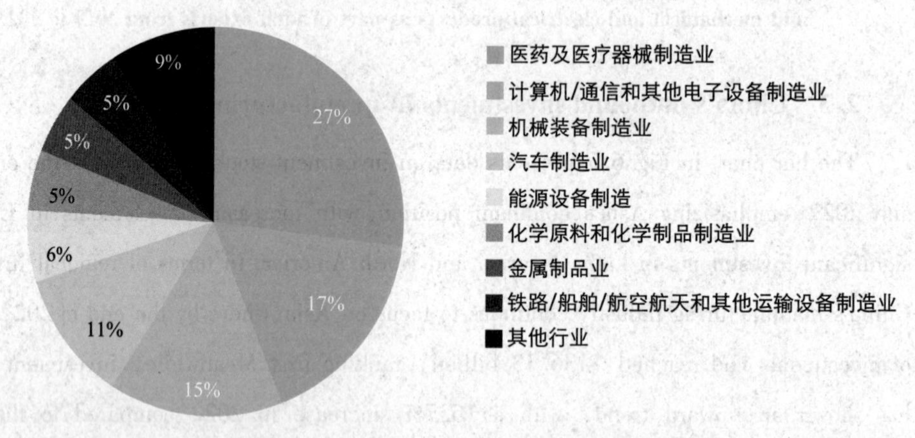

注：数据下载于2024年2月8日，近三年为2021—2023年，统计数据包括交易状态为“完成”和“待完成”的交易；部分交易未披露金额；目的地不包括中国香港地区。

近三年中国制造业海外并购投资交易数量产业部门分布情况

Fig. 6. 8 Distribution of overseas M&A transactions in China's manufacturing industry by sector (2021 – 2022)

in the pharmaceutical and medical device manufacturing industry, computer, communication, and other electronic equipment manufacturing, mechanical equipment manufacturing, and automotive manufacturing. The manufacturing industry of railways, ships, aerospace and other transportation equipment, the metal products industry, and the chemical raw materials and chemical products manufacturing industry all account for the lowest percentage of 5%.

Task 3 *Complete the following analysis based on Task 1 and Task 2.*

3.1 China is a global manufacturing leader (Fig. 6.9)

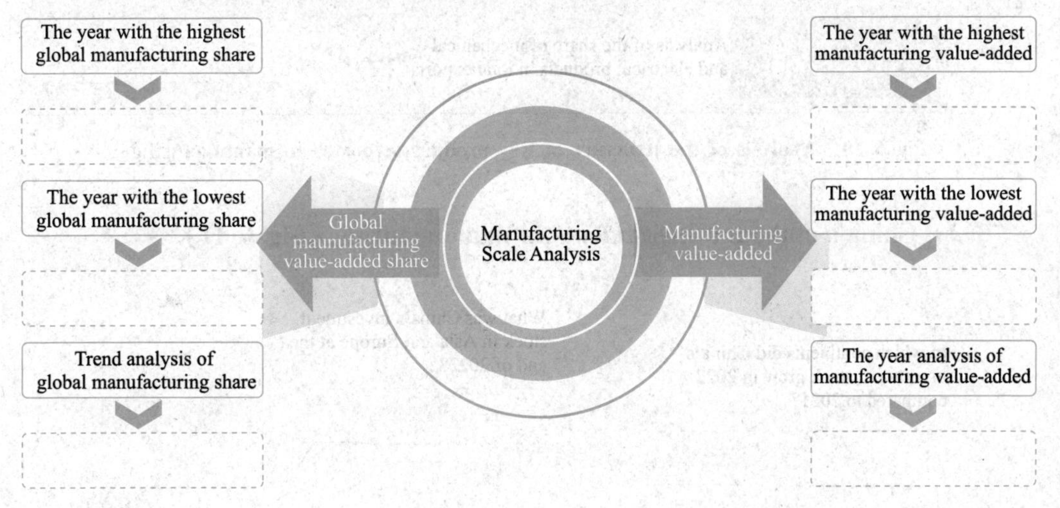

Fig. 6.9 Manufacturing scale analysis

3.2 Structure of key products exported in manufacturing (Fig. 6.10)

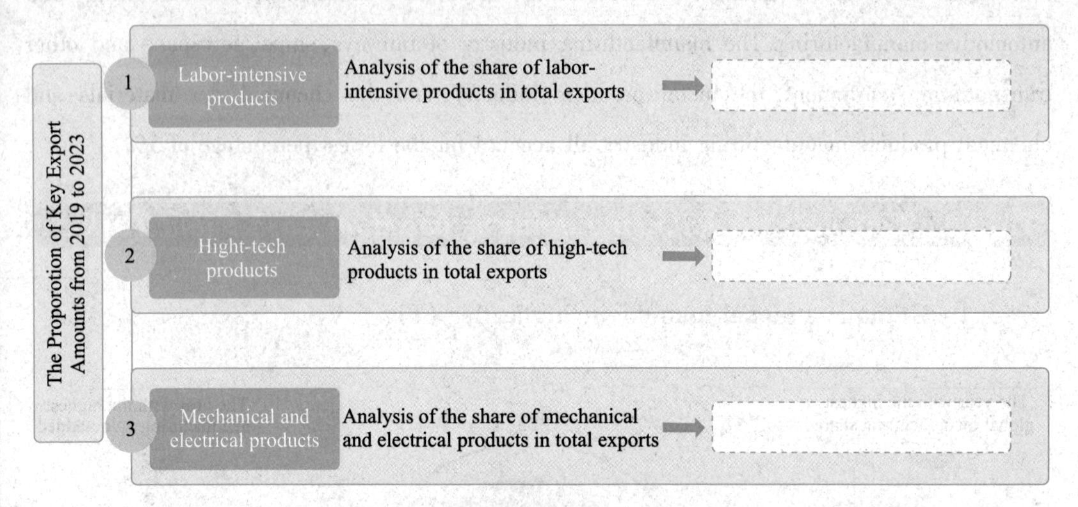

Fig. 6.10 Analysis of the structure of Rey products exported in manufacturing

3.3 China's outbound investments in manufacturing (Fig. 6.11)

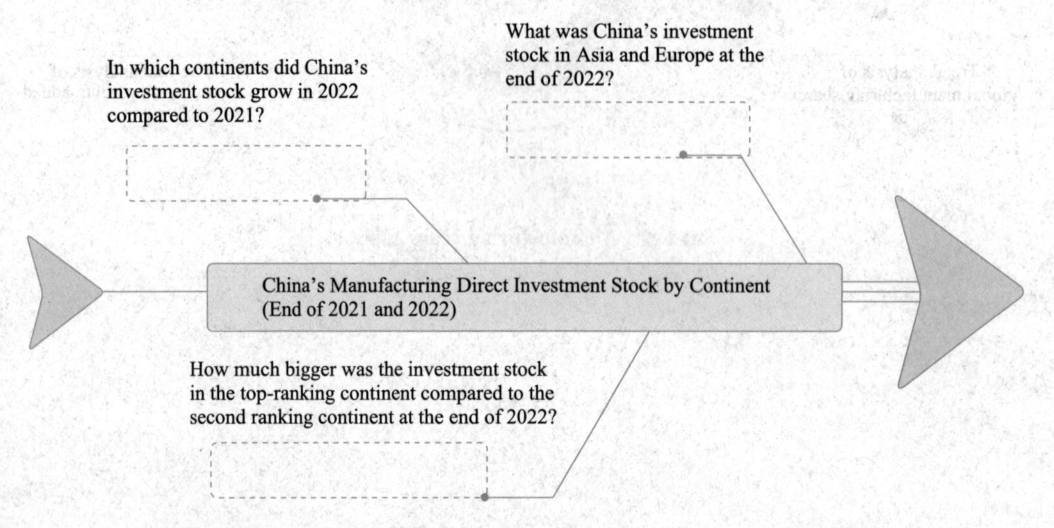

Fig. 6.11 Analysis of China's outbound investment in manufacturing

3.4 Distribution of overseas M&A investments in manufacturing (Fig. 6.12)

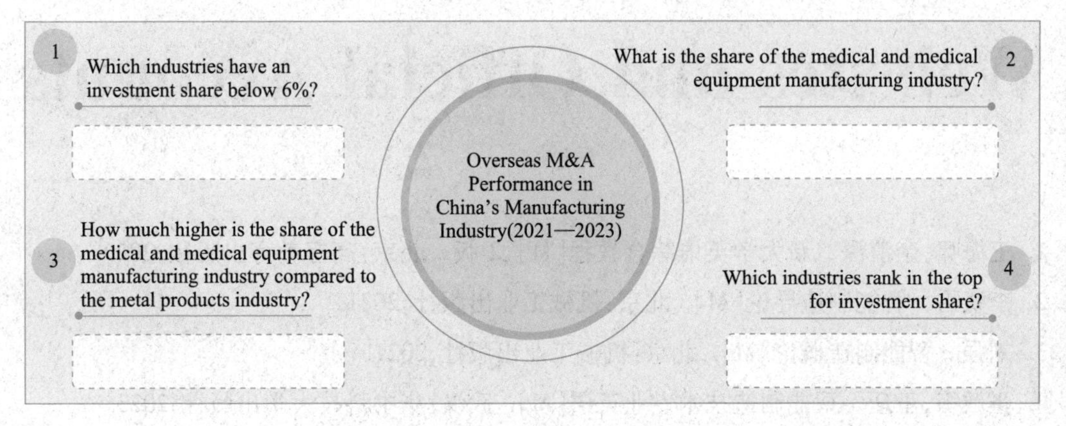

Fig. 6.12 Distribution of overseas M&A investments in manufacturing

Appendix: Reference, Websites and Official Accounts

1. 王海啸,余渭深. 新大学英语综合教程[M],2 版. 北京:高等教育出版社,2022.
2. 李晓雪. 智能制造导论[M]. 北京:机械工业出版社,2021.
3. 刘强. 智能制造概论[M]. 北京:机械工业出版社,2021.
4. 张敬衡,王珏. 智能制造技术专业英语[M]. 武汉:华中科技大学出版社,2023.
5. CHEN Y. Integrated and Intelligent Manufacturing:Perspectives and Enablers [J]. Engineering, 2017(3):588 - 595.
6. https://www.chinadaily.com.cn/
7. 《中国制造业企业出海白皮书》, 毕马威, kpmg.com/cn/socialmedia, 2024.
8. 艾瑞. "2022 年中国智能机器人行业研究报告". 艾瑞咨询, https://mp.weixin.qq.com/s/znKgCkPPzNJBYFP8IDA8Xg, 2022.
9. 艾瑞. "2023 年中国工业互联网平台研究报告". 艾瑞咨询, https://mp.weixin.qq.com/s/oVsmZu2gqb-pSkstbnGpDg, 2024.
10. 先进制造 AMC. "走进灯塔工厂 (12): 宁德时代". 先进制造 AMC, https://mp.weixin.qq.com/s/ogD2PUZmTIdnX3cR7JIUNw, 2024.
11. 廖芮言. "工业数字孪生系统中六轴机械手臂虚实联动". 木棉树软件, https://mp.weixin.qq.com/s/YO_ts-kV811xaeduWstF8Q, 2024.
12. LIKY. "人工智能 + 智慧制造" 之创新案例篇. "AIDT 智能工业, https://mp.weixin.qq.com/s/UTopcNjw5xmR5A8K6kVUTA, 2024.
13. e-works. "压轴重头戏! 2021 中国工业数字化转型领航企业 50 强出炉". 数字化企业, https://mp.weixin.qq.com/s/zGHd-OyY5xwJ39cKXJC1dA, 2022.